Motorcycle Engineering

Motorcycle Engineering is a primer and technical introduction for anyone interested in motorcycles, motorcycling, and the motorcycle industry. It provides insight into how motorcycles are made and operated.

Motorcycles, mopeds, and scooters are important factors in world transport, and they are playing an increasingly important role in transport policy as we move towards greater environmental awareness. Motorcycles and scooters give freedom of personal transport that enable large commuter distances to be covered quickly and easily. Their small footprint offers easy storage, as only minimal space is required. To celebrate the importance of motorcycles on the world stage, a brief history is included with a detailed timeline detailing the development of the motorcycle alongside major world events.

Written in an accessible fashion, no previous knowledge of engineering or technology is required, as all technical terms are readily explained and a glossary and abbreviation list is included. Whether you are an enthusiast, racer, student, or industry professional, you will surely find this an enjoyable read and a handy reference book on your shelf.

Andrew Livesey, MA, CEng, is a lecturer in engineering at Ashford College University Centre in Kent, England, when he is not riding his motorcycle, bicycle, or building something in his garage. He is a member of several motorcycle and bicycle groups, mixing social riding with competition events when possible. He also enjoys very fast high-performance cars. His Routledge publications include *Basic Motorsport Engineering* (2011), *Advanced Motorsport Engineering* (2012), *The Repair of Vehicle Bodies, 7th edition* (2018), *Practical Motorsport Engineering* (2018), and *Bicycle Engineering and Technology* (2021).

Motorcycle Engineering

Andrew Livesey

LONDON AND NEW YORK

First published 2021
by Routledge
2 Park Square, Milton Park, Abingdon, Oxon OX14 4RN

and by Routledge
52 Vanderbilt Avenue, New York, NY 10017

Routledge is an imprint of the Taylor & Francis Group, an informa business

© 2021 Andrew Livesey

The right of Andrew Livesey to be identified as author of this work has been asserted by him in accordance with sections 77 and 78 of the Copyright, Designs and Patents Act 1988.

All rights reserved. No part of this book may be reprinted or reproduced or utilised in any form or by any electronic, mechanical, or other means, now known or hereafter invented, including photocopying and recording, or in any information storage or retrieval system, without permission in writing from the publishers.

Trademark notice: Product or corporate names may be trademarks or registered trademarks, and are used only for identification and explanation without intent to infringe.

British Library Cataloguing-in-Publication Data
A catalogue record for this book is available from the British Library

Library of Congress Cataloging-in-Publication Data
Names: Livesey, Andrew, author.
Title: Motorcycle engineering / Andrew Livesey.
Description: Abingdon, Oxon ; New York, NY : Routledge, 2021. | Includes index. | Summary: "The complex engineering behind the machines is explained in easy-to-understand terms and supported by 350 images. It covers a range of motorcycle types, and will be particularly useful for students on motorcycle and motorsport courses such as those run by the IMI and City & Guilds, as well as BTEC programmes"— Provided by publisher.
Identifiers: LCCN 2021000799 (print) | LCCN 2021000800 (ebook) | ISBN 9780367419202 (hbk) | ISBN 9780367419196 (pbk) | ISBN 9780367816858 (ebk)
Subjects: LCSH: Motorcycles—Design and construction.
Classification: LCC TL440 .L57 2021 (print) | LCC TL440 (ebook) | DDC 629.227/5—dc23
LC record available at https://lccn.loc.gov/2021000799
LC ebook record available at https://lccn.loc.gov/2021000800

ISBN: 978-0-367-41920-2 (hbk)
ISBN: 978-0-367-41919-6 (pbk)
ISBN: 978-0-367-81685-8 (ebk)

Typeset in Sabon
by KnowledgeWorks Global Ltd.

Contents

Abbreviations and symbols   xi
General abbreviations   xiii
Preface   xvii

1 Power unit – engine   1

Identification 2
Engine performance 2
Engine construction 8
Combustion 25
Cylinder head 27
Short block assembly 29
Induction 31
Exhaust 32

2 Ignition and fuel   35

Ignition system 35
Battery 36
Ignition switch 36
Ignition coil 37
Fuel system 45
Air:petrol ratio 50
Petrol injection 51
Electronic control unit 52

3 Lubrication and cooling   55

Lubrication system 55
The cooling system 65

4 Health, safety, security, and the environment — 75

Personal health and safety procedures 75
COSHH 86
RIDDOR 86
Miscellaneous topics 93

5 Motorcycle types — 95

Motorcycle categories, ages, and licence requirements in the UK 95
Safety equipment 96

6 Materials for motorcycles — 111

Metallic materials 111
Manufacture of steel 112
Annealing and tempering 113
Classifications of steel 113
Alloying metals used with steel in motorcycle frames 115
Work hardening and fatigue failure 117
Properties of materials 117
Frame tubing 118
Reynolds tubing 118
Columbus tubing 120

7 Frames and fairings — 123

Frames 123
Fairings 125

8 Electric motorcycles — 133

Rules and regulations 133
Electric bicycle construction 136
Health and safety 142

9 Soldering, brazing, and welding — 143

Comparison of fusion and nonfusion jointing processes 143
Aluminium brazing 149
Health and safety and the environment 152
Safety measures 154
General equipment safety 156
Gas-shielded arc welding (MIG, MAG, and TIG) 156

10 Running gear and braking — 159

Running gear 159
Suspension and steering 159
Wheels and tires 166
Wheels 168
Tires 169
Braking system 176
Mechanical brakes 178
Hydraulic brakes 181
Drum brakes 184

11 Motorcycle electrical and electronic systems — 197

Battery 197
Alternator 201
Starter motor 202
Vehicle circuits 203
Lights 209

12 The motorcycle industry — 213

Traditional motorcycle shop 213
Assembly methods: modern retailing 218
Other motorcycle industry services 220
Trade associations 222

13 Reinforced composite materials — 225

Introduction 225
Basic principles of reinforced composite materials 226
Manufacture of reinforced composite materials 227
Types of reinforcing material 228
Resins used in reinforced composite materials 230
Catalysts and accelerators 231
Pre-impregnated material (Pre-preg) 236

14 Data — 239

The norm 239
Understanding and using data 240

Bar charts and stacking bar charts 242
Pie charts 242
Normal distribution 243
Profit and loss account (PLA) 245

15 Science terminology 249

SI system and common units 249
Decimals and zeros 250
Accurate measuring 251
Capacity and volume 253
Temperature and heat 253
Force and pressure 254
Amps, volts, ohms, watts, and kirchhoff 255
Friction 256
Some common laws of mechanical engineering 257
Impact and momentum 257

16 Transmission 259

Clutch 259
Gearbox 264

17 Tuning and customizing 275

Modifications 275
Vehicle-fixed coordinate system 293
Earth-fixed coordinate system 293

18 Inspection, test, and rebuild 299

Safety first 299
Test and rebuild 302
Costs 304
Health and safety and the environment 306

19 Sustainability 325

The three centric concerns of sustainability 325
Transforming our world: the 2030 agenda for sustainable development 327
Discussion 328
Insurance 329

20 Brief history of motorcycles	331
Purpose and use 334	
World war I 335	
Between the wars 339	
World war II 340	
The golden era 340	
1969 – The sea change 342	
The millennium 344	
Timeline 345	
Appendix – Apprenticeship standard for a motorcycle technician	351
Glossary	355
Index	359

Abbreviations and symbols

The abbreviations are generally defined by being written in full when the relevant technical term is first used in the book. In a very small number of cases, an abbreviation may be used for two separate purposes, usually because the general concept is the same but the use of a superscript or subscript would be unnecessarily cumbersome; in these cases, the definition should be clear from the context of the abbreviation. The units used are those of the internationally accepted *Systemé International* (SI). However, because of the large American participation in motorcycling and the desire to retain the well-known imperial system of units by many enthusiasts, where appropriate imperial equivalents of SI units are given. Therefore, the following is intended to be useful for reference only and is neither exhaustive nor definitive.

Greek alphabet symbols

- α (alpha) angle – tire slip angle
- λ (lambda) angle of inclination
- μ (mu) coefficient of friction
- ω (omega) rotational velocity
- ρ (rho) air density
- η (eta) efficiency
- θ (theta) angle

General abbreviations

a	acceleration
A	area – frontal area of motorcycle and rider; or ampere
ABS	antilock braking system; or acrylonitrile butadiene styrene (a plastic)
AC	alternating current
AF	across flats – bolt head size
AFFF	aqueous film forming foam (firefighting)
bar	atmospheric pressure – 101.3 kPa or 14.7 psi as standard or normal
BATNEEC	best available technique not enabling excessive cost
BS	British Standard
BSI	British Standards Institute
C	Celsius or Centigrade
CAD	computer-aided design
CAE	computer-aided engineering
CAM	computer aided manufacturing
C_D	aerodynamic coefficient of drag
CG	center of gravity, also CoG
CIM	computer integrated manufacturing
C_L	aerodynamic coefficient of lift
cm	centimeter
cm^3	cubic centimeters – capacity; also called cc. 1000 cc is 1 liter
CO	carbon monoxide
CO_2	carbon dioxide
COSHH	Control of Substances Hazardous to Health (Regulations)
CP	center of pressure
CR	compression ratio
D	diameter
d	distance
dB	decibel (noise measurement)
DC	direct current
deg	degree (angle or temperature), also °

dia.	diameter
DTI	dial test indicator
EC	European Community
ECU	electronic control unit
EFI	Electronic Fuel Injection
EN	European Norm – European Standard
EPA	Environmental Protection Act; or Environmental Protection Agency
EU	European Union
f	frequency
F	Fahrenheit, force
ft	foot
ft/min	feet per minute
FWD	front-wheel drive
g	gravity; or gram
gal	gallon (U.S. gallon is 0.8 of UK gallon)
GRP	glass-reinforced plastic
HASAWA	Health and Safety at Work Act
HGV	heavy goods vehicle (used also to mean LGV – large goods vehicle)
HP	horsepower (CV in French, PS in German)
HSE	Health and Safety Executive; also, health, safety, and environment
HT	high tension
I	inertia
ID	internal diameter
IMechE	Institution of Mechanical Engineers
IMI	Institute of the Motor Industry
in^3	cubic inches – measure of capacity
IR	infrared
ISO	International Standards Organization
k	radius of gyration
kph	kilometers per hour
l	length
L	wheelbase
LH	left hand
LHD	left-hand drive
LHThd	left-hand thread
LPG	liquid petroleum gas
lumen	light energy radiated per second per unit solid angle by a uniform point source of 1 candela intensity
lux	unit of illumination equal to 1 lumen/m^2
M	mass
MAX	maximum
MIG	metal inert gas (welding)

MIN	minimum
N	Newton; or normal force
Nm	Newton meter (torque)
No	number
OD	outside diameter
OL	overall length
OW	overall width
P	power, pressure, or effort
Part no	part number
PPE	personal protective equipment
pt	pint (UK 20 fluid ounces, USA 16 fluid ounces)
PVA	polyvinyl acetate
PVC	polyvinyl chloride
Q	heat energy
r	radius
R	reaction
Ref	reference
RH	right hand
rpm	revolutions per minute; also RPM and rev/min
RTA	Road Traffic Act
RWD	rear-wheel drive
std	standard
STP	standard temperature and pressure
TE	tractive effort
TIG	tungsten inert gas (welding)
V	velocity; or volt
VOC	volatile organic compounds
W	weight
w	width
WB	wheel base
x	longitudinal axis of vehicle or forward direction
y	lateral direction (out of right side of vehicle)
z	vertical direction relative to vehicle

Superscripts and subscripts are used to differentiate specific concepts.

S I UNITS

cm	centimeter
K	Kelvin (absolute temperature)
kg	kilogram (approx. 2.25 lb)
km	kilometer (approx. 0.625 mile or 1 mile is approx. 1.6 km)
kPa	kilopascal (100 kPa is approx. 15 psi, that is atmospheric pressure of 1 bar)
kV	kilovolt

kW	kilowatt
l	liter (approx. 1.7 pint)
l/100 km	liters per 100 kilometers (fuel consumption)
m	meter (approx. 39 inches)
mg	milligram
ml	milliliter
mm	millimeter (1 inch is approx. 25 mm)
N	newton (unit of force)
Pa	pascal
ug	microgram

IMPERIAL UNITS

ft	foot (= 12 inches)
hp	horse power (33,000 ft lb/minute; approx. 746 watt)
in	inch (approx. 25 mm)
lb/in^2	pressure, sometimes written psi
lb ft	torque (10 lb ft is approx. 13.5 Nm)

Preface

Author at Whitstable Charity Motorcycle Event

Motorcycling gives you a freedom to travel. You are at one with the open road and can park in the smallest of spaces. Pressing the start button on my motorcycle and seeing the tachometer needle rise makes me feel ready to lap the world. Lots of riders are in fact doing just that on their adventure motorcycles, riding the silk routes across the Himalayas or the plains and jungles of Africa.

Motorcycles are very reliable now, so the need for maintenance is a minimum. This book is written as a primer to help you learn about the motorcycle, its operation, and the motorcycle business. It is hoped that it might encourage you to make motorcycles, or at least customize or modify your motorcycle. Or maybe it will tempt you into the motorcycle industry in some way. Motorcycle sales are increasing, and the range of available machines is amazing. Thank you to all the motorcyclists who have helped me on this motorcycle journey, especially 21st Moto Ltd of Swanley, Kent.

I hope that it will give you a pleasurable and informative read, to enable you to enjoy motorcycling as much as I do.

Eur Ing Andrew Livesey, MA, CEng, MIMechE, FIMI
Herne Bay, Kent
Andrew@Livesey.US

Chapter 1

Power unit – engine

'There ain't no substitute for cubes.'

Whether it's cubic inches or cubic centimeters, the more of them that you have in your engine, the more power you can develop. That's how it is for petrol and diesel engines – with electric motors, you need volts and amps.

Motorcycle sport is often grouped into engine sizes, so the competitor is challenged to get the most power out of the engine. There are also usually regulations on what is, and is not, allowed to improve the power output.

Tech Note

The term **horsepower** – **HP** – comes from steam engine sales agents of about 200 years ago saying how many horses their engines could replace. In French, this is **cheval vapour (CV)**; in German, this is **pferde starke (PS)**.

One HP, in any language, that is CV or PS, is equal to 33,000 foot-pounds per minute, that is 746 watts.

When we are talking about power output, we must be careful to compare like for like. When we say **BHP** (brake horsepower), we are talking about the engine power measured on an engine brake or dynamometer – **dyno**. But rolling road dynamometers measure power at the wheels – this is after the frictional losses in the transmission system – typically around 10% to 15%. Also, there are a number of different standards for measuring power outputs set by different organizations. The most popular are the American **SAE** standard and the German **DIN** standard. Both measure power but under different conditions, with variations like the use of air filters and the way that the cooling system is connected.

Power is about doing work in an amount of time – mathematically, it's work done per unit time.

2 Motorcycle Engineering

Figure 1.1 Kick start to get the engine going on a lightweight motorcycle.

IDENTIFICATION

Identification of the engine before working on it is very important. The **VIN** number will help identify the type, or classification, of the engine. The detail of the engine will be given in a separate engine number, the pre-fix will identify the engine type, and the serial number will identify the exact engine.

ENGINE PERFORMANCE

The two common terms used are:
  Power – this is **work done** in unit time.
  Torque – turning moment about a point.
  Let's discuss them for clarity and then look at the calculations. When we are using the term *power*, we are referring to how much energy that

the engine has. A big heavy motorcycle needs a big powerful engine. Power is about doing work in a time period; it means burning fuel in the time period. We can make a small four-cylinder engine – say one from a motorcycle like a Kawasaki ZX6R – produce over 100 BHP from its 600 cc, but we need it to rev at about 12,000 rpm.

For a mathematical definition of these terms, we need to start with work done. Work done is the amount of load carried multiplied by the distance travelled. The load is converted into force: for instance, the force needed to move the motorcycle in newtons (N). The distance is measured in meters. That is:

$$\text{Work done}(Nm) = \text{Force}(N) * \text{Distance}(m)$$

As we also express torque in Nm, it is common to use the term joule (J) for work done.

Tech Note

Joule is a term for energy. 1 J = 1 Nm

If we use a force of 10,000 N to take a drag-bike down a 200-m drag strip, then we have exerted 2,000,000 Nm, or 2,000,000 J. We'd say 2 mega joules (2 MJ). We'd need to get this amount of energy out of the fuel that we were using

The force is generated by the pressure of the burning gas on top of the piston multiplied by the area of the top of the piston. So, the work done is the mean (average) force of pushing the piston down the cylinder bore multiplied by the distance traveled.

Example

The work done during the power stoke of an engine where the stoke is 60mm and the mean force is 5kN

$$\begin{aligned}\text{Work Done} &= \text{Force}*\text{Distance}\\ &= 5 \text{ kN}*60 \text{ mm}\\ &= 5000N \times 0.06m\\ &= 300 \text{ J}\end{aligned}$$

Tech Note

The mathematical symbols used in this book are those found on your calculator or mobile phone:
  * is multiply and / is divide

The same mean force is going to create the torque. This time we are going to use the crankshaft throw – this is half the length of the stoke.

Example

Using the same engine

$$\text{Torque} = \text{Force} * \text{Radius}$$
$$= 5 \text{ kN} * 30 \text{ mm}$$
$$= 5000 N * 0.03 m$$
$$= 150 \text{ Nm}$$

The work done by a torque for one revolution is the mean force multiplied by the circumference. The circumference is 2Πr so:

$$\text{WorkDone} = F * 2\Pi r$$

$$\text{As} Fr = T$$

So, Work Done = $2\Pi T$

That is for one revolution. For any number of revolutions, where n is any number, the formula is:

Work Done in n revolutions = $2\Pi n T$

Example

Using the same engine of the previous examples:
The work done in 1 minute at 6000 rpm will be:

$$\text{WD in } n \text{ revolutions} = 2\Pi n T$$
$$= 2 * \Pi * 6000 * 150$$
$$= 5657 \text{ kJ}$$

Power is, as we said, work done in unit time, which is:

Power = Work Done / Time

The motorcycle industry uses a number of different units and standards for power. From our calculations we can use watts (W) and kilowatts (kW) and then convert.

Tech Note

1 kW = 1000 W

1 watt = 1 J/second
1 kW = 1 kJ/s

Example

Following on from our engine in the previous calculations and examples:

Power = Work Done / Time
 = 5657 kJ / 60
 = 94.3 kW

The term **horsepower** (HP or hp) was derived by James Watt as the average power of a pit pony. These were small horses used to turn pulleys to draw water from Cornish tin mines (pits) before steam power became more popular. He equated the power of his steam engines to a number of these pit ponies. For our purposes 1 HP equals 33,000 ft-lb/minute.

In French horsepower is cheval vapour (CV); in German, it is pferde starke (PS).

For conversion purposes, 1 HP is equal to 746 W.

When talking about power and doing work, the weight of the bike comes into play. It's worth doing a comparison with performance motorcycles to get a good picture.

TERMINOLOGY

One metric ton is 1000 kg. As a kilogram is equivalent to 2.25 pounds, a metric ton is the equivalent of an imperial ton – 2250 lb.

Table 1.1 Typical BHP per ton figures

Bike/car	Capacity	BHP	Weight (kg)	BHP per ton
Kawasaki ZX6R	599 cc	130	185	702
Harley Davidson	883 cc	53	263	201
Triumph Bonneville T120	1198 cc	80	224	357
Suzuki Hayabusa	1299 cc	173	251	689
Kawasaki H2 motorcycle	1000 cc supercharged	310	215	1442
Typical BTCC car	2-liter turbo	350	1000	350
German Touring Motorcycle - DTM	4-liter supercharged	500	1122	445
Bugatti 16/4	8 liter with 4 turbos	1200	1990	603

The aim of any motorcycle designer is to get the maximum BHP per ton, not forgetting to have enough torque to get off the starting line – especially for hill climb events.

Fuel energy output measurement

The energy output of fuel varies. There are two ways of measuring the quantity of fuel: by the kilogram or by the liter. Energy companies tend to give outputs for kilograms. This is a more stable method of measurement and is traditionally used for ships and airplanes where mass is a more important factor. Typical values for consumer fuels are:

- Petrol 45.8 MJ/kg
- Diesel 45.5 MJ/kg

However, diesel is about 15% more dense than petrol. When measured at 15.55 °C this figure is used, as it is the equivalent of 60 °F – the temperature scale used in non-European countries. The specific gravity (also called relative density) of consumer fuels is:

- Petrol 0.739
- Diesel 0.82 to 0.95

The range of specific gravity (sg) values for diesel is because of the variety of products for commercial use and summer and winter products. Winter diesel is lighter than summer diesel to prevent waxing. When diesel turns solid – referred to as waxing – it takes place at between 14 °C and 18 °C, with the latter temperature for winter use.

So, taking typical consumer fuels – pump fuels – we get the following values per liter:

- Petrol 33.7 MJ/liter
- Diesel 36.9 MJ/liter

Tech Note

There have been a variety of diesel motorcycles over the years, mainly for military use. As we are in a period of change, the author has included it for interest and to highlight its good qualities.

Alternative fuels – Certain classes of racing allow, sometimes insist on, the use of other fuels. **Methanol** is a clean-burning fuel; it is used in the United States. **Biodiesel** like methanol is made from renewables and is therefore environmentally friendly. **Liquified natural gas** (LNG) is a product of

Table 1.2 Typical energy outputs of alternative fuels

Fuel	MJ/kg	MJ/liter
Methanol	19.9	15.9
Biodiesel	37.8	33.3
LNG	38	25.5

shale – bored or fracked from the ground – and is not environmentally friendly. There are various forms of gas products – they all burn dry, and because of their dry burning characteristics limit engine life, even with special operating procedures. LNG/ Liquid Petroleum Gas (LPG) is used by cost-conscious operators, as it attracts a lower level of duty or tax.

Bomb calorimeter – This is used to measure the energy output of fuel. It is shaped like you would imagine a bomb to look and works by burning – exploding – a spoon full of fuel to see how much energy it will generate. You do this by measuring the temperature change brought about by burning the fuel under controlled conditions. Bomb calorimeters are readily available and allow you to check the fuel that you are using. Remember that for maximum power you want the fuel which produces the highest energy output for a given volume, so it must be high-density, high-calorific output.

A typical calculation using a bomb calorimeter

The formula:

$$\begin{aligned}\text{Calorific value} &= \text{Total water mass} \times \text{Temperature rise} \\ &\quad \times \text{Specific heat of water} \\ &= 2.652 * 4.58 * 4.19 \\ &= 50{,}900 \text{ kJ}/\text{kg}\end{aligned}$$

Table 1.3 Data for bomb calorimeter

No.	Name	Reading
1	Density of fuel – density	0.845
2	Water equivalent of calorimeter – mfg figures	652 g
3	Mass of water	200 g
4	Mass of fuel	1 g
5	Temp. of water before combustion	19.26 °C
6	Temp. of water after combustion	23.84 °C
7	Temperature increase, that is No. 6–No. 5	4.58 °C
8	Specific heat of water	4.19 kJ/kg deg K

ENGINE CONSTRUCTION

> ### Key points
>
> - The engine burns fuel at a very high temperature to force the piston down the cylinder bore.
> - The connecting rod and crankshaft convert the reciprocating motion of the piston into rotary motion.
> - Engine size is calculated as a product of the cylinder bore and the piston stroke.
> - Engines may be petrol (SI) or diesel (CI).
> - Two-stroke petrol engines can be found in some small motorcycles and garden machinery.

Almost all motorcycle engines are petrol engines. Royal Enfield and one or two other manufacturers have made diesel motorcycle engines; but these were mainly for military use.

> ### Tech Note
>
> Race engines are developed to produce the maximum power with a specific capacity to conform to the regulations of the racing classification – this usually means high engine speed with a narrow power band.

Cylinder block and crankcase

There are quite a lot of different designs of motorcycle engines, as most of the manufacturers have what they consider to be their ideal design. In the car industry many manufacturers share an engine; for instance, you will find Ford engines in Peugeot, Renault, and Fiat. In low-volume-production motorcycles, you will find the use of other manufacturers' engines. You will also find the same engine used across a range of motorcycles from the same manufacturer but at different levels of tune.

To reduce weight and improve reliability, motorcycle engines and gearboxes are usually made in what is referred to as unit construction: this uses one main casting for the engine crankcase and gearbox housing. This term is often misused and confused with other concepts. Most front-wheel-drive motorcycles use a variation on this concept. Early engines were made from cast iron; almost universally motorcycle engines are mainly aluminum alloy. The crankcase and gearbox housing are one main casting; on top of this is bolted the cylinder block, and on top of the cylinder block is bolted the cylinder head.

If you consider this you can see that you can use one crankcase casting to serve a whole range of engine options:

- The block can be changed within the same fitting to the crankcase to give different sizes of bore and stroke, that is, different engine capacity options.
- The cylinder heads and valve arrangements can be changed to give different levels of tune – that is, different performance levels.

Pistons – These move up and down in the cylinder bores. This up-and-down movement is called **reciprocating motion**. The piston forms a gastight seal between the combustion chamber and the crankcase. The burning of the fuel and air mixture in the combustion chamber forces the piston down the cylinder to do useful work. The pistons are usually made from aluminum alloy for its lightweight and excellent heat-conducting ability. The top of the piston is called the crown; the lower part is called the skirt. The pistons must be perfectly round to give a good seal in the bore when the engine is at its normal running temperature. However, aluminum expands a lot when it is heated up. The pistons have slits in their skirts to allow for their expansion in diameter from cold to their normal operating temperature. When cold, pistons may be a slightly oval shape, so that when at running temperature they are a perfect fit in the cylinder bore.

The pistons are fitted with piston rings to ensure a gastight seal between the piston and the cylinder walls. This is needed to keep the burning gases in the combustion chamber. The piston rings are made from close grain cast iron, a metal that is very brittle. But the piston rings are slightly springy because of their shape. Usually there are three piston rings. The top two are compression rings to keep the gases in the combustion chamber. The bottom one is an oil ring; its job is to scrape the oil off the cylinder walls. The oil is returned to the sump by passing through the slots in the piston rings and running down inside the pistons. The piston rings are made from cast iron – this is very brittle, so when piston rings are being fitted great care must be taken not to break them.

Tech Note

Pistons on race engines are as light as possible to give very high-speed running. To reduce their weight, the skirts are made very short and cut away to a *slipper* shape – referred to as slipper pistons.

> **Tech Note**
>
> What do the terms bore and stroke mean?
>
> The bore is the cylindrical hole, or cylinder, in which the piston runs. The bore must be perfectly smooth, round, and parallel. It is also the term used to describe the diameter of the cylinder; this is usually expressed in millimeters (mm) or inches (in). The stroke is the distance the piston travels from the bottom of the cylinder – called bottom dead center (BDC) – to the top of the cylinder – called top dead center (TDC). The stroke also may be measured in millimeters or inches. The surface inside the cylinder is called the cylinder wall.

Connecting rod (con rod) – This connects the piston to the crankshaft. The con rod has two bearings: the **little end** connects to the piston and the **big end** to the crankshaft. The con rod is made from either cast iron or forged steel. The big end bearing is a shell bearing; this allows for easy replacement and cheap manufacture.

Crankshaft – This, in conjunction with the con rod, converts the reciprocating motion of the pistons into the rotary motions, which turn the flywheel. The crankshaft is located in the cylinder block by the main bearings. The big end bearings are attached to the crank pins; the crank pins are at the ends of the crank webs. The distance between the center of the **crank pin** and the center of the **main bearing** is called the **throw**. The throw is half of the **stroke**.

> **Key points**
>
> From TDC to BDC is the stroke. For the piston to travel from TDC to BDC, the crankshaft rotates 180 degrees – half a revolution. The crank pin is moved from above the main bearing – the length of one throw – to below the main bearing – the length of another throw. That is, two throws are equal to the length of the stroke.

Cylinder head (head) – The head sits on top of the cylinder block. The head contains the combustion chambers and valves. Between the head and the block is a cylinder head gasket. The cylinder head gasket allows for the irregularities between the block and the head and keeps a **gastight seal** for the combustion chamber. The engine cylinder head locates the spark plugs.

Valve cover and sump – The cylinder head is fitted with a valve cover (also called a **rocker box** or **cam box**). The valve cover encloses the valves and their operating mechanism, forming an oil-tight seal for the engine oil. The bottom of the crankcase on some engines is fitted with a sump.

The sump has two jobs; it is a store for the engine lubricating oil and forms an oil-tight seal to the bottom of the engine. Both the valve cover and the sump are usually made from thin pressed steel. The sump is removed to gain access to the oil pump and bearings.

Timing mechanism – At the front or side of the engine is the timing mechanism. This is a belt, a chain, or a shaft that connects the crankshaft to the camshaft. A casing covers the timing mechanism. The timing end of the engine is often called the free end. **Cylinder numbers** always start from the free end. On some engines, the chain is in the middle of the engine between the cylinders.

Flywheel – The flywheel is attached to the crankshaft. The flywheel end of the engine is the drive end. That is, the flywheel turns the clutch and the gearbox to move the vehicle.

Tech Note

As a mechanic, it is good to understand the different metals that are used in engines. The different metals must be considered when you are handling the part, and particularly when tightening up nuts and bolts.

- Cast iron is used for cylinders and cylinder heads on older engines; it is very heavy and brittle, so it will break if dropped.
- Aluminum is lightweight and expands a great amount when heated up. It is also soft, so it is easily scratched. You must be careful not to over-tighten spark plugs in aluminum cylinder heads or you will damage the threads.
- Pressed steel is used for sumps and valve covers; this is easily bent. A bent sump may leak around the joints.
- Hardened steel is used for the crankshaft; this is both heavy and expensive.

Four-stroke petrol engine

The four-stroke petrol engine works on a cycle of four up-and-down movements of the piston. These up-and-down movements are called strokes. The piston moves down from TDC to BDC, then up to TDC again. Each stroke corresponds to half of a turn of the crankshaft; therefore, the complete cycle of four strokes takes two revolutions of the crankshaft.

The petrol and air mixture is burned in the combustion chamber during one of the strokes. The heat from the burning fuel causes a pressure increase in the combustion chamber. This pressure forces the piston down the bore to do the useful work. The mixture is ignited by the spark plug – hence the term spark ignition (SI).

12 Motorcycle Engineering

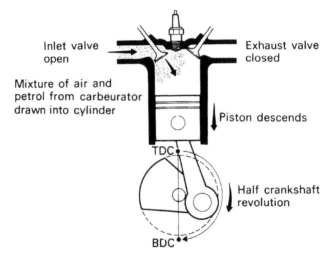

a *Induction stroke*

Figure 1.2 Induction stroke.

The cylinder head is fitted with inlet valves; these open and close to control the flow of the petrol and air mixture from the inlet manifold into the combustion chamber. The cylinder head is also fitted with exhaust valves to control the flow of the spent exhaust from the combustion chamber into the exhaust manifold and exhaust system. The passage in the cylinder head, which connects the manifold to the combustion chamber, is called the port. There are inlet ports and exhaust ports. The valves are situated where the

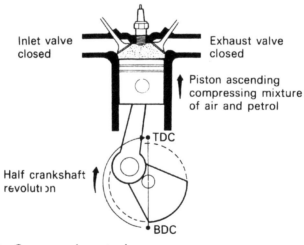

b *Compression stroke*

Figure 1.3 Compression stroke.

Power unit – engine 13

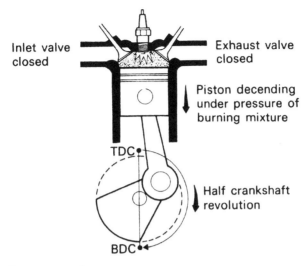

c *Power stroke*

Figure 1.4 Power stroke.

ports connect into the combustion chamber. The valves are operated by the camshaft; this is discussed later in this chapter.

Induction stroke

The piston travels down the cylinder bore from the TDC, drawing in the mixture of petrol and air from the inlet manifold. This is like a syringe

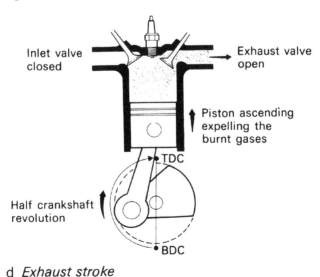

d *Exhaust stroke*

Figure 1.5 Exhaust stroke.

drawing up a liquid. The downward movement of the piston has caused a depression above the piston. This depression, or partial vacuum, is satisfied by the air coming into the inlet manifold through the air filter. The air mixes with the petrol that is supplied from either from the injectors or a carburetor.

Compression stroke

When the piston reaches BDC, it starts to return up the bore. At about BDC the inlet valve is closed by the camshaft; the exhaust valve was already closed. The mixture of petrol and air, which was drawn in on the induction stroke, is now compressed into the combustion chamber. This increases the pressure of the mixture to about 1250 kPa (180 psi). The actual pressure depends on the compression ratio of the engine – on race engines, it is typically between 10:1 and 16:1. On WSB and BSB machines, this figure may exceed 20:1. Increasing the compression ratio within limits will increase the power output.

Nomenclature

The mathematical sign : means to and signifies a ratio.

Power stroke

As the piston reaches TDC on the compression stroke, the spark occurs at the spark plug. This spark, which is more than 40 kV (40,000 volts), ignites the petrol–air mixture. The mixture burns at a temperature of over 2000 degrees Celsius and raises the pressure in the combustion chamber to over 5000 kPa (750 psi). The pressure of the burning petrol–air mixture now starts to force the piston back down the cylinder bore to do useful work. The piston rings seal the pressure of the burning mixture into the combustion chamber, so that the pressure exerts a force on the piston, the gudgeon pin, and then the con rod, which converts this downward motion into rotary motion at the crankshaft. It's good to remember that a modest-speed engine will fire on each cylinder over 50 times every second.

Tech Note

How much force does the burning mixture actually exert on the con rod or crankshaft, and why do they not bend?

The amount of force depends on the size of the engine; but as a rough guide, imagine an elephant sat on the top of the piston every time it goes down. The components will not bend as long as the engine is rotating and the force is being passed on to the transmission to move the vehicle.

Exhaust stroke

At the end of the power stroke the exhaust valve opens. When the piston starts to ascend on the exhaust stroke, this is the last stroke in the cycle; the burnt mixture is forced out into the exhaust. The mixture of petrol and air has been burnt to change its composition. Its energy has been spent. The temperature of the exhaust gas is about 800 to 1200 degrees Celsius. The petrol–air mixture has been burnt to become carbon monoxide (CO), carbon dioxide (CO_2), water (H_2O), nitrogen (N), and free carbon (C). The exhaust gas is passed through the exhaust system to the catalytic converter to be cleaned and made nontoxic. The exhaust gas exits the engine in waves – it is not a continuous stream, nor in parcels like a sausage machine. The speed of sound at 20 °C is 343 m/s – the different exhausts give different pitches; compare the exhaust noise of a Ferrari V12 with a Ducati twin.

Flywheel inertia

There is only one firing stroke for each cycle. The flywheel keeps the engine turning between firing strokes. Single-cylinder engines need a bigger flywheel in proportion to their size than those with more cylinders do. The flywheel on a large V8 engine is smaller than one on a four-cylinder engine. The flywheel's desire to keep rotating is called inertia; it is inertia of motion.

Valve operation

The inlet and the exhaust valves each open once every two revolutions of the crankshaft. The mechanism for opening the valves is a camshaft, which is either direct acting on the valves or operates them through a push rod and a rocker shaft assembly.

If the camshaft is situated above the valves, the engine is referred to as overhead cam (OHC). Other types are overhead valve (OHV) and side valve. A rubber-toothed belt, a chain, or a bevel gear shaft may drive the camshaft. As the camshaft must rotate at half the speed of the crankshaft to open the valves once for every two revolutions of the crankshaft, the drive is through a 2:1 gear ratio. The cam gear wheel has twice as many teeth as the crankshaft gear wheel.

The actual point at which the valves open and close depends on the engine design. The motorcycle's workshop manual will give the valve timing figures; these are usually expressed in degrees of the crankshaft relative to TDC and BDC.

It is essential that the valves close firmly against their seats to give a good gastight seal and allow heat to conduct from them to the cylinder head so that they may cool down. The valves are held closed by springs. To ensure that the valves close firmly, even when the components have expanded because of the heat, the valve mechanism is given a small amount of clearance. The valve clearance is measured with a feeler gauge; a typical figure is 0.15 mm (0.006 in). If the valve clearance is too great, then a light metallic rattling noise will be heard.

Two-stroke petrol engine

The two-stroke petrol engine is used mainly in small motorcycles, although they have been used in some larger racing machines. It operates on one up-stroke and one down-stroke of the piston; that is, one revolution of the crankshaft. The most common type is the Clerk Cycle engine; this has no valves, just three ports. The three ports are the inlet port, the transfer port, and the exhaust port. The flow of the gas through these ports is controlled by the position of the piston. When the piston is at TDC, both the transfer and the exhaust ports are closed. When the piston is at BDC, the piston skirt closes the inlet port.

The piston travels up the bore. As it reaches TDC, it closes both the transfer and the exhaust ports. At the same time, the piston is compressing the

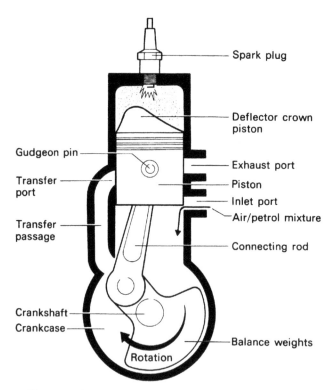

a *Two-stroke petrol engine: piston approaching TDC*

Figure 1.6 Two-stroke approaching TDC.

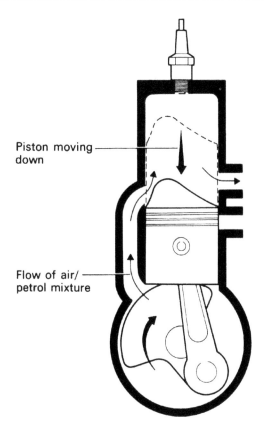

b *Two-stroke petrol engine: piston approaching BDC*

Figure 1.7 Two-stroke approaching BDC.

charge of petrol and air above it into the combustion chamber. At about TDC the spark plug ignites the petrol–air mixture. The burning of the petrol–air mixture increases the temperature and the pressure so that the burning gas pushes the piston down the bore. The downward force of the piston passes through the gudgeon pin to the con rod and crankshaft to drive the vehicle.

While the piston is ascending, its skirt uncovers the inlet port. The upward motion of the piston causes a vacuum in the crankcase, which is satisfied by the petrol–air mixture from the carburetor entering through the now open inlet port.

Tech Note

Two-stroke engines have a separate crankcase that is not combined with the gearbox.

When the piston is traveling downwards – being forced down by the burning mixture – the piston crown first uncovers the exhaust port. This allows the spent gas to escape into the exhaust system. The skirt of the piston covers the inlet port at the same time as the piston crown uncovers the transfer port at the top of the cylinder. The underside of the piston therefore acts like a pump plunger, forcing the fresh charge of petrol and air that is in the crankcase up and through the transfer port into the cylinder so that another cycle is started.

The two-stroke petrol engine is much lighter and simpler than the four-stroke petrol engine and has fewer moving parts. It has neither valves nor a valve operating mechanism. Two-stroke petrol engines usually run at high speeds. The problem is that oil must be mixed with the petrol for lubrication – this causes the exhaust to smoke.

Tech Note

Two-stroke engines were used in the early motorcycles, and there are racing classes for these; however, because of emission regulations, no new ones are being made. Old racing two-stroke motorcycles were loved by many for the smell of the burnt oil – even though it could be a health hazard. The oil was vegetable based and called *Castrol R* – it is currently available for historic applications.

Four-stroke diesel engine

The operation of the four-stroke diesel engine is similar to the four-stroke petrol engine. The diesel engine draws in air only and then compresses this at a very high pressure and temperature to cause the fuel to combust and burn. The fuel is injected directly into the engine at a very high pressure. Vehicles with diesel engines are economical, and they produce lots of pulling power at low engine speeds. Some military motorcycles have used diesel engines, as their advantage is that other than for starting, an electrical supply is not needed. This is an advantage in that it can be operated without any electrical power supply and is therefore unstoppable by military equipment that can cut out ignition systems.

Flywheel inertia

Because of their high compression ratios and heavy moving parts, diesel engines usually have large flywheels.

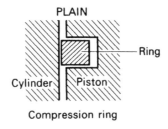

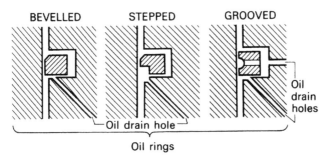

Types of piston rings

Figure 1.8 Types of piston rings.

Firing order

By increasing the number of cylinders, the engine becomes more compact in terms of size and smoother running. Smoothness of running is further improved by setting the sequence in which the cylinders fire; this is called the firing order. The normal firing orders for four-cylinder engines are 1-3-4-2 and 1-2-4-3.

Engine capacity

To find the capacity of an engine, first you need to find the size of each cylinder; this is called the swept volume. This is the volume that is displaced when the piston goes from BDC to TDC. The swept volume of each cylinder is the product multiplying the cross-sectional area and the stroke.

$$\text{Swept Volume} = \Pi D^2 L / 4$$

$\Pi = 3.142$
$D = $ Diameter of bore
$L = $ Length of stroke
All divided by 4

The engine capacity is the product of the swept volume and the number of cylinders.

20 Motorcycle Engineering

Capacity = Swept Volume * Number of Cylinders

You will find the following abbreviations:
V_s = Swept Volume
N = Number of Cylinders
The engine capacity is usually measured in cubic centimeters (cc).

NOMENCLATURE

There are 1000 cc in 1 liter. A 1000-cc motorcycle is therefore referred to as 1 liter. American motorcycles are sized in cubic inches (cu in). One liter is equal to 62.5 cu in.

Compression ratio

The compression ratio is the relationship between the volume of gas above the piston at BDC compared to that at TDC. You need to know the swept volume (V_s) and the volume of the combustion chamber, which is referred to in this case as the clearance volume (V_c).

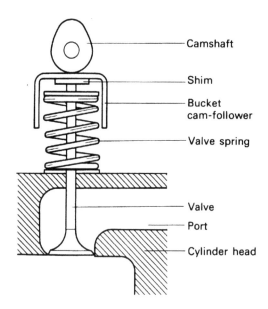

Overhead cam layout

Figure 1.9 OHC valve layout.

Power unit – engine 21

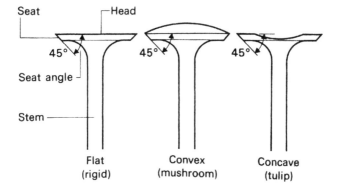

The three types of valves most commonly used

Figure 1.10 Types of valves.

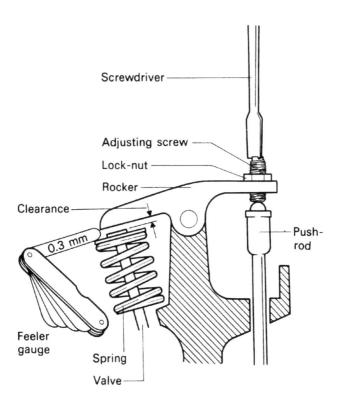

Valve-clearance adjustment

Figure 1.11 OHV clearance being adjusted.

Compression Ratio = $(V_s + V_c) / V_c$

Road-version petrol engines have compression ratios of between about 9:1 and 11:1. Race engines are up to about 20:1. Diesel engines are usually up to about 22:1.

Volumetric efficiency – In other words, the efficiency of the engine in getting fuel and air to fill the cylinder. The fact that an engine has cylinders of 1 liter (61 cu in) does not mean that you are getting that amount of air and fuel into the cylinders. The flow of gas is affected by a number of points, mainly:

- Size, shape, and number of valves
- Valve timing

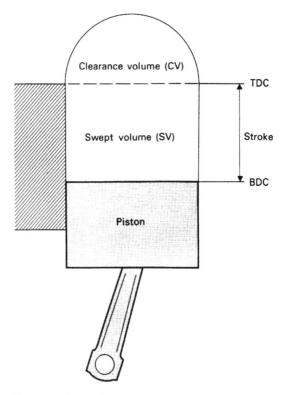

Compression ratio

Figure 1.12 Compression ratio.

- Size and shape of the inlet and exhaust ports
- Shape and location of the combustion chamber
- Bore-to-stroke ratio
- Engine speed
- Type of induction tract – air filter, carburetor or throttle body, inlet manifold
- Type of exhaust system – manifold, silencers, cat, and pipe layout

The volumetric efficiency is calculated by measuring the amount of air entering the engine and finding this as a percentage of the actual volume of the engine. The formula is:

Volumetric Efficiency = Actual Air Flow / Capacity of Engine

The air flow can be measured on the test dynamometer by using either an air flow meter on the inlet or calculating the air flow from the pressure drop across an orifice. You would normally do this over a set time period so that the air flow is calculated in liters per minute (cu in/minute or cu ft/minute).

A typical engine might have a volumetric efficiency of 80%. A well-tuned naturally aspirated engine may have an efficiency approaching 90% with throttle bodies and all the right manifolds. With a turbo charger or supercharger fitted that is pressurizing the air into the cylinder, the figure will be over 130%.

The greater the amount of air that can be packed into the cylinder, the more power the engine is going to give out. The saying is, *there ain't no substitute for cubes*. That is the more cubic inches (liters) of air, the more power the engine will develop. An example of an affordable engine with a high volumetric efficiency is the Kawasaki HS2 – this gets 310 BHP from 1 liter.

Thermal efficiency – From volumetric efficiency you saw that you need to get the engine efficient in terms of getting the air (actually air and fuel) in to and out of the engine. That is one thing. Now when the air and fuel are in the engine they need to be burned efficiently to produce as much energy as possible to be turned into power, and that conversion of burning gas into turning the crankshaft needs to be as efficient as possible to get the maximum power out of the engine.

Thermal Efficiency = Power Output / Energy Equivalent of Power Input

The power output is easy to measure on a dynamometer, either as an engine in a test rig or the complete vehicle on the rolling road.

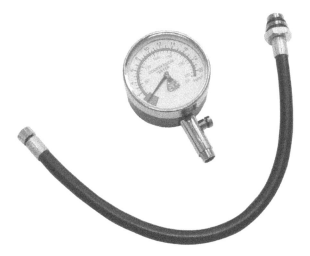

Figure 1.13 Compression gauge.

> **Tech Note**
>
> Lots of drivers know the power output of their engines; very few know the efficiency of the same engine – greater efficiency means fewer refueling stops and hence faster race times; in other words, more wins!

To calculate the energy input, you are going to need to know:

- Mass of air used – The volume of air multiplied by the density. The density depends on altitude, temperature, and weather conditions.
- Mass of fuel
- Specific calorific value of the fuel – How much energy it produces for a given amount.

The calculation for the mass of air follows on from the measurements of air flow from the volumetric efficiency calculations. The calculations for density may be made by either sampling or using tables. When carrying out calculations or comparing figures, it is worth checking whether they are at STP or NTP.

STP

STP – standard temperature and pressure – is defined as air at 0 °C (32 °F) and 1 bar (14.7 psi).

In the imperial and U.S. system of units, it is defined as air at 60 °F and 14.696 psi.

NTP

NTP – normal temperature and pressure – is defined as air at 20 °C (68 °F) and 1 bar (14.7 psi).

Air density at sea level at NTP is 1.2 kg/m³ (0.075 lb/ft³).

Tech Note

Have you ever noticed that on a cold, damp day the engine runs well and sounds well? This is because the air is at maximum density and the water cools the cylinder to prevent detonation.

The energy from fuel is discussed under fuel composition separately in this chapter; examples of calculations are also given.

Tech Note

Some simple changes, like chipping, may increase the power by 5% or 10% but increase fuel consumption by 40%. Customers may have a shock at the increase in fuel costs.

Valve timing – This is very important on an engine. The basic principles are to open the valves as quickly as possible, at as great a distance possible, and for as long as possible. The problem is that if you do this, the engine will need to run fast to maintain gas velocity, so it is likely to stall at low engine speeds, or at least produce low power. To combat this, many vehicles use variable valve timing.

If you fit a high-performance camshaft, you need to ensure that the timing is exact – especially on engines where fully open valves may touch the piston crown. To this end, a **vernier adjustable timing pulley** may be needed.

Cam design – There are several methods of cam design. These are **constant velocity, constant acceleration**, and **simple harmonic motion**.

COMBUSTION

Flame travel – The combustion of the air and fuel mixture is a burning action – we often refer to it as an explosion because it is very fast; however, the fuel burns with a flame, which is traveling very quickly. At 6000 rpm each cylinder fires 50 times each second; in a Formula 1 engine at 22,000 rpm this is 183 times per second. In this time the piston has completed four strokes, so the stroke time is a quarter of this – that is, the maximum duration of the burning if it were between TDC and

BDC would be 0.005 s (5 ms) at 6000 rpm and 0.0016 s (1.6 ms) at 22,000 rpm. On race engines, the spark plug needs to be as near to the center of the combustion chamber as possible to give even burning of the gas. The faster the engine speed, the more important this is. Flame travel should not be confused with spark plug burn time, which is the length of time that the plug is sparking.

Pre-ignition, post-ignition, pinking, and **detonation** are often confused, as their symptoms and effects are almost identical; that is, the air and fuel are burned in a noisy manner, usually a knocking noise, and there is a loss of power or uneven running. **Running-on** is associated with these symptoms too. Pre-ignition is when the fuel is ignited before the spark occurs; this is usually by something burning or becoming overheated in the combustion chamber. Often this is the spark plug insulator, a carbon deposit, or a section of damaged cylinder head gasket. Post-ignition is when the fuel burns late, or more likely, combustion continues after the engine is switched off. When the engine continues to run for a while when the ignition is switched off, this is called running-on. Pinking, also called **knock**, is the noise made by the mixture burning too quickly or on two flame fronts and is caused by the ignition timing being too advanced or too low-octane-rated fuel being used. Detonation is when pockets of fuel burn in an irregular way, usually because of poor air–fuel mixing, incorrect ignition timing, or poor combustion chamber design.

Octane rating – There are two methods of measuring fuel antiknock ability:

- **Research Octane Number** (RON): A measure of antiknock during acceleration under medium load
- **Motor Octane Number** (MON): A measure of antiknock during acceleration under heavy load

As neither are perfect measures, an average of the two is often used; this is called the **antiknock index**.

Antiknock Index = (RON + MON) / 2

Cetane rating –The diesel equivalent of octane rating.

SAFETY NOTE

Petrol is highly flammable – race petrol is especially volatile – so please take care when handling or storing any petrol.

Figure 1.14 Chrome-plated cylinder head.

CYLINDER HEAD

Valve layout – The valve layout for high-performance engines is designed to get the maximum amount of air and fuel into the engine and then out again as exhaust gas. The most common design uses four valves. Because the inlet valves are larger than the exhaust valves, the exhaust gas is forced out by the piston under high pressure. The gas flows from one side of the engine to the other to reduce the time taken; this allows an increase in engine speed.

Cooling – For good cooling, aluminum cylinder heads are almost universal – they dissipate the heat very quickly. Coolant (water) jackets may be opened out to improve coolant flow.

Valves – These use a variety of materials to give longer valve life and better cooling. Typically, high chromium alloy steel is used to resist corrosion and give wear resistance. Valve head shapes tend to be then – often called a penny on a stick – for minimum weight and less mass for heat buildup. Variations of seat angles and seat materials are used to give good sealing

---

Tech Note

The thinner the valve seat, the tighter the seal; conversely, the wider the seat, the better the heat transfer from the valve to cylinder head.

With aluminum alloy cylinder heads, seat inserts are needed to give the required hardness; iron cylinder heads may also have inserts for competition use – these are often fitted at the same time that the cylinder head is machined to fit larger valves. Valve seat inserts may be screwed (now rare) or press-fit into the cylinder head.

To improve thermal efficiency, two factors of combustion chamber design should be considered: **swirl ratio** and **surface-to-volume ratio**. The formulas are quite self-explanatory:

Swirl Ratio = air rotation speed / crankshaft speed

The higher this figure, the better the engine will run – look for figures over 5.

Surface to volume ratio = Surface area of combustion chamber / volume of combustion chamber

The lower this number, the better the engine will run – a sphere will give the best ratio.

Valve springs – Double- or triple-valve springs may be used to enable high revs.

Camshaft – When building a competition cylinder head, you should pay attention to the fit of the camshaft(s). If possible, use white metal bearings and line bore to ensure perfect straightness.

Tech Note

Turn the camshaft like the crankshaft when setting it up; that is, ensure that it turns freely at each step of tightening the caps or when inserted – you will need to do this before fitting the valves.

Standard camshafts tend to be cast iron or low-grade steel – for competition use, you will need a high-grade alloy steel. These can be hardened by **nitriding**. Lower-quality competition camshafts may be **Tockle hardened, Tuffrided**, or simply coated to give a better wearing surface.

Camshaft drive – For competition use, high-tensile bolts will be used.

Gas flow – This is a commonly misused term or used as a catch-all for anything to do with cleaning and polishing cylinder heads. The volume and velocity of gas flow through a cylinder head are measured on a **flow bench**. The cylinder head is attached to the flow bench so that the air flowing through it can be measured. The exhaust valve is kept firmly shut while the inlet valve is opened in small increments and the flow recorded. The same can then be done with the exhaust valve. The modifications are then made to the relevant part of the cylinder head – typically, the inlet port is smoothed and polished, then the flow is measured again. It is not the absolute flow rate which is important – this may be difficult to measure in a finite way – the percentage improvement is the indicator looked for. A 5% improvement is very good. The factors affecting gas flow are:

- Port shape
- Port finish
- Manifold to port fit – do they align?
- Gasket fit – does it overlap the port?
- Valve seat angle
- Valve size
- Valve opening

SHORT BLOCK ASSEMBLY

Materials – For lightness and cooling dissipation, aluminum alloy is the obvious choice – but this will depend on the racing class. To improve aluminum block strength and improve wear resistance, **cryogenic treatment** is often carried out – check the regulations to see if it is allowed for the class. This involves taking the temperature of the block down to a temperature of about –180 degrees C (–300 degrees F) for a period of about 48 hours. This smooths out the grain structure of the metal, making it stronger.

> **Tech Note**
>
> You can cryogenically treat any part of the motorcycle – brake discs (rotors) and pads are a common and economical choice.

Liners are used in aluminum blocks to give the essential resistance to wear. They may be **dry liners** or **wet liners**. Wet liners are the most common to give easy-to-make open blocks. If you are cytogenetically treating the liners, you will need to bore them out about 0.002 inch (two thou) to return them to true round – you must bear this in mind if you are blueprinting the engine.

> **Tech Note**
>
> Never hit liners, even with something soft – they easily distort.

Drillings – When building an engine, check the drillings and passageways to ensure that they are clear and the correct size to give sufficient water and oil flow.

Crankshaft – This is made from high-grade alloy steel with a high nickel content so that it can be nitride. This involves leaving it in a bath of ammonia at 500 degrees C (900 degrees F) for about 24 hours.

When fitting the crankshaft, ensure that it turns after each bearing cap is tightened. A very small amount of very light (5 SAE) oil should be placed on each journal and bearing shell, but ensure that there is no oil behind the shell. Also check for crankshaft end float after fitting the thrust bearings. End float must be the minimum to ensure con-rod alignment and clean clutch operation – remember the clutch thrusts against the crankshaft thrust bearings.

Con rods – These are increasingly made from carbon fiber composite, as this method allows for relatively low-cost batch production. The alternatives are high-strength aluminum and titanium. Consider the speed at which they are traveling, and you will appreciate the need for aerodynamics in their design. Changing the engine stroke or the piston height may mean changing the con rods.

Bore–stroke ratio – This is simply the ratio of the two dimensions:

$$= \text{bore} / \text{stroke (mm or inches)}$$

If the bore is larger than the stroke, the engine is said to be an oversquare engine; if bore and stroke are the same, it is said to be a square engine; if the stroke is greater than the bore, then it is called a long stroke engine. Generally, race engines are oversquare and rev at high speeds – this can be over 20,000 rpm. Older engines tend to be long stroke, being lower running and producing more torque than ultimate power.

Balancing – Pound (dollar) per HP, balancing an engine gives the best return. Not only does it ensure smooth operation and longer life, it also allows higher revs at the top end. Balancing breaks down into two parts: **static** and **dynamic**.

Static balancing can be done on the bench with a simple weight scale. Dynamic balancing needs a specialized machine. To carry out static balancing, you start with the pistons. Weigh each piston and remove material from the heavier ones until they all weigh the same. Reduce the weight of the heavy pistons by removing material from the bottom inside of the skirt or a relatively unstressed part of the boss.

Tech Note

If you have access to more than one box of pistons of the same size, weigh them all – you may find the number that you want that are all the same to save machining.

Next are the con rods. Support the little end on a piece of wire (or similar) and weigh the big ends on the scale. Do the same, supporting the big end. Remove metal by drilling small holes until they are all the same weight.

Weight can be added to steel rods by drilling small holes and filling them with lead – make sure that they are very clean and use flux so that there is no risk of the filler coming out.

You can statically balance the flywheel and clutch by assembling them onto the crankshaft, supporting the crankshaft in vee blocks and spinning, then wait until it stops. The heavy part will be at the bottom. Again, remove weight by drilling small holes in unstressed parts. When it is balanced, you should be able to stop the assembly at any point without it rotating.

When properly statically balanced, dynamic balance should be easy. Indeed, if on a budget, static balance only will give very good results.

Surface finish – Surface finish is important for mating surfaces and surfaces that are subject to air or other fluids flowing over them. On a race engine, good surface finish makes the engine look good – a very important point. Nicely polished parts make the engine attractive.

Blueprinting – This is another often misused term. It simply means building the engine to the manufacturer's specification – in other words, the blueprint – or drawing. Final designs were printed on large sheets of blue paper. As we now use computer-aided drafting (CAD), this means we no longer print drawings in blue. When blueprinting, you should pay detailed attention to each part fit and ensure that it is the correct size within design tolerance.

INDUCTION

Gas laws – Air is not always what you think it is. You may think that air goes through the engine like sausages through a sausage machine. You may think that a chunk of it goes into the cylinder to be processed and then it comes out as exhaust gas. Certainly, air is mixed with petrol; it does go into the cylinder; it is burnt and it does come out as exhaust. However, the gas is traveling at several hundred miles per hour (hundreds of kilometers per hour). And it travels in waves, not in sausage-like lumps.

Induction systems fall into three categories:

- **Open induction**, usually the carburetor, or the throttle body has an open ram pipe
- **Closed induction**, this uses a plenum chamber with air filter
- **Forced induction**, as with a turbocharger or supercharger

The **ram pipe** length can be changed to suit the engine operating speed – this is very much a trial-and-error activity. The plenum (Latin for full) chamber is designed to give the engine a supply of still air. With forced induction, air is in effect pumped into the cylinder under pressure. The ram pipes guide the air into the cylinder and help control the waves.

> **Tech Note**
>
> Changing from a plenum chamber to ram pipes can add 5% to the power output of some engines.

EXHAUST

Exhaust gas travels in the same way – that is, it comes out in waves. The waves are generated by the piston in the same way as, but opposite to, the induction waves. The piston going up the bore pushes the gas out, not as a sausage, but as a pulse. The pulse velocity depends on a number of factors which we will look at later in this section. First of all, let us look at the speed of sound. Exhaust, as you know, makes a sound, so the pulse wave is traveling at the speed of sound. The speed of sound varies with a number of variables.

> **HISTORIC RACER NOTE**
>
> Race bikes at Brooklands had to have their exhaust tail pipe extend past the rear wheel spindle to satisfy the complaining residents of nearby St. Georges Hill – this led to fancy fishtail silencers.

The **speed of sound** in air (C) is calculated from the heat capacity (γ) of air (1.4), the air pressure (p), and the air density (ρ).

$$C = \sqrt{\gamma p / \rho}$$

Example

At NTP:

$$C = \sqrt{1.4 * 101000 / 1.2} = 343 m/s$$

The speed of sound varies with temperature (T) in the ratio:

$$C_2 / C_1 = \sqrt{T_2 / T_1}$$

Taking T1 as NTP of 20 degrees Celsius and C1 as 343 m/s and remembering that this should all be in absolute temperature form Kelvin (K), at 800 degrees C the speed of sound will be:

$$C_2 / 343 = \sqrt{800 + 273 / 273 + 20}$$
$$= 656 m/s$$

Now we have the speed of sound in the very hot exhaust. This allows us to work out the length of the exhaust **primary pipe** to suit the engine speed. You'll see lots of variations of exhaust manifolds and systems. The primary pipe length from the exhaust valve to the 'Y' joint into the rest of the system or the open pipe on a dragster is important to give maximum power at any particular speed. The reason for this is that as the exhaust valve opens, it sends a pulse wave at the speed of sound – we calculated that – down the primary pipe. When it gets to the end of the primary pipe and the gas can expand, the sound wave travels back down the pipe to the exhaust valve; this then returns back to the end of the primary pipe.

Tech Box

Try this
 Think of the exhaust gas as a wave on the sea shore – it goes in and out even though the tide is coming in all the while.

If you can get the primary pipe to be the length that will be just right for gas to go up and down between the exhaust valve opening and closing, it will leave a negative pressure (vacuum) at the exhaust valve as it closes. This will scavenge (clean out) the exhaust gas from the engine most efficiently. So, let's look at how we do this. We know that time taken to cover a distance is that distance divided by the velocity:

$\text{Time}(t) = \text{distance}(d) / \text{velocity}(v)$

The distance is from the exhaust valve to the end of the primary pipe and back – we'll call this 2L. The velocity is the speed of sound (C). The time varies with engine speed and valve period (the number of degrees the exhaust valve is open, for example, 120 degrees). One hundred twenty degrees is 0.333 of a revolution (360 degrees). Six thousand rpm equates to 100 rev/sec, so 0.333 of a revolution will take 0.00333 seconds, that is: 0.00333/100 = 0.00333. We calculated that the speed of sound at 800 degrees C is 656 m/s. So the distance traveled by the sound wave at 656 m/s in 0.00333 s is:

$0.00333 * 656 = 2.18 m$

The length of the primary pipe is half of this – remember it is down and back, so it is 1.09 m.

Tech Note

The same principle applies to the induction, which is why you see old racing motorcycles with long inlet manifolds and long ram pipes.

Helmholtz theory – If you look at the inner tube in a silencer, you see that it is drilled with lots of holes. Helmholtz worked out that if you chop the end off a wave, it will reduce its energy – in other words, the noise that it makes. So as the sound wave passes each hole in a silencer, a part of it goes through the hole, reducing the energy and hence the noise.

SAFETY NOTE

Exhaust gas is both hot and poisonous.

Chapter 2

Ignition and fuel

The ignition system and the fuel system are separate systems, but they work closely together, almost as one.

IGNITION SYSTEM

The purpose of the ignition system is to provide a spark in the combustion chamber of the petrol engine, which will ignite the mixture of petrol and air while it is under pressure. As the piston compresses the petrol–air mixture on the compression stroke, the pressure may be increased to over 2000 kPa (300 psi). The voltage needed for the spark to jump across the spark plug gap at this high pressure is about 40 kV (40,000 volts).

There are many different types of ignition systems. The ones that you are likely to come across on motorcycles are:

1. Magneto system – also called direct ignition
2. Coil ignition – also called battery ignition or Kettering system
3. Capacitor discharge ignition – an early form of electronic ignition
4. Digital electronic ignition using an electronic control unit (ECU)

Magneto – This is amazingly simple and used on small scooters, mopeds, lawnmowers, and chainsaws. The flywheel contains a number of magnets – these spin around when the engine is turning. On the engine crank case is a series of coils – these are stationary so that when the magnets spin over the stationary coils, electricity is generated. Also on the crankcase is a set of contact breaker (cb) points. The cam on the flywheel opens and closes the cb points. This is set up so that when the cb points start to open, the piston is at a point ready for the plug to spark. The coils are connected to the spark plugs with high-tension leads. The ignition timing can be adjusted by moving the position of the cb points.

The spark is strong when the engine is turning at moderate speed. It is weak (low voltage) at low and high speeds. This means, of course, that the kick-start mechanism needs to turn the engine over fairly quickly to get a reasonable spark – if the mixture is not exact, this makes for difficult starting. It also means the spark plugs and plug leads need to be well maintained – it is normal to carry a spare spark plug and plug wrench.

Coil ignition – The original system, which has been used for about 100 years, is the Kettering system, designed by Dr. Kettering. The fully electronic and other systems are easier to understand if you first understand the mechanical-electrical Kettering system. The main components of the Kettering system are the **battery, coil, spark plugs**, and **distributor**. Let's look at these components individually and see how the system works.

Tech Note

Lots of classic motorcycles use coil ignition. It is fairly standard on machines fitted with carburetors up to about the 2000 model year.

BATTERY

The battery is the **source of electrical power** for the ignition system (and other systems). It is usually located in the engine compartment. The battery supplies electrical power to the ignition switch. The battery has a nominal voltage of **12 volts**; the ignition circuit draws about 0.5 amp. The battery is usually connected on the negative side (–) to the chassis earth and on the positive side (+) to the main **fuse box**.

RACER NOTE

Wiring on race bikes is usually through a separate isolator switch and may incorporate a connection for a trolley-mounted starter battery or power pack.

SAFETY NOTE

Battery acid is corrosive – mind that you do not spill it.

IGNITION SWITCH

This is connected to the battery through the main fuse box. The ignition switch makes and breaks the ignition circuit. When the engine is not running, the switch must be off to disconnect the electrical power supply and prevent the coil from overheating. Switching the ignition off also switches off other related circuits and prevents the battery from being discharged. The ignition switch is also a security device to prevent the vehicle from being stolen, are used, unlawfully. The ignition switch is combined with the steering lock for added security and the convenience of needing only one key for both the steering lock and the ignition switch. The ignition switch/steering lock is6 situated at the top of the steering tube. Most motorcycles also have a kill switch to stop the engine in an emergency.

IGNITION COIL

The ignition coil is a kind of transformer. It changes the **low tension** (**LT**) 12 V from the battery to a **high tension** (**HT**) 10 kV at the spark plugs.

> ### NOMENCLATURE
>
> LT refers to the components of the ignition system that operate at a nominal voltage of 12 V. HT is that part of the ignition system that operates at several thousand volts. The HT can range from 5 kV for an older bikes to 40 kV for current models. We refer to the battery has having a nominal voltage of 12 V. We say this is a 12 V battery, but the actual voltage may be between 9 V and 16 V, depending on a number of factors. A kilovolt is 1000 volts.

The coil has three electrical terminals: one that goes to the ignition switch, one to the distributor body, and one to the distributor cap. The ones to the ignition switch and the distributor body are LT; the one to the distributor cap is HT.

> ### SAFETY NOTE
>
> The high-voltage spark at the spark plugs, and associated leads, can kill you. If the system is wet or you touch a bare cable or component, the shock will cause you to react and you may hit your head, or elbow, on another part of the motorcycle, which could result in a serious injury.

The connection to the ignition switch is the 12 V power supply. The connection to the distributor body goes to the contact breaker points, which carry out the switching action. The HT to the distributor provides the spark for the spark plugs.

The ignition coil operates on the principle of **difference of turns**. That is, it has two separate **windings** wound around a **soft iron core**. The **primary winding** is connected to the distributor LT; the **secondary winding** is attached to the distributor HT. The secondary winding has many more turns of wire than the primary winding; the increase in voltage from LT to HT is proportional to the difference in the number of turns of the wire.

Ignition coils are usually filled with oil to improve cooling. Be careful not to damage the coil case, as this is usually soft aluminum and if damaged may allow the oil to drain and the coil to overheat.

Spark plugs

The metal end, or **body**, of the spark plug screws into the cylinder head so that the **electrode** protrudes into the combustion chamber. The screw thread with the washer and mating surfaces form a gastight seal, thus preventing the loss of compression from the cylinder.

The spark plug has two electrodes: a **center electrode** and a **side electrode**. The terminal end unscrews, on some motorcycles the plug lead is attached with an eyelet, on others it is push-on as is common on cars.

The diameter and the reach of the spark plug vary from engine to engine. The most common diameters are 10 mm, 14 mm, and 18 mm. The common reaches are 3/8 inch, ½ inch, and ¾ inch. It is important that the correct reach of spark plug is fitted to the engine; otherwise, the electrode may foul against the piston crown. The diameter will only fit the tapping size in the cylinder head. The spark plug can be identified by the letter and number code that is printed on either the **insulator** or the body.

Tech Note

How can I check that one make of spark plug matches another?

Spark plugs are made by a number of different companies, for example: Champion, Bosch, and NGK. Most motor factors and accessory shops have lists of equivalents. These lists show the various makes and models of motorcycles and tabulate the code numbers for the different spark plug manufacturers.

When spark plugs are replaced, typically at between 16,000 km (10,000 m) for old vehicles and 120,000 km (80,000 m) for modern vehicles, they must be replaced with the correctly coded ones. The spark plug manufacturer's

Ignition and fuel 39

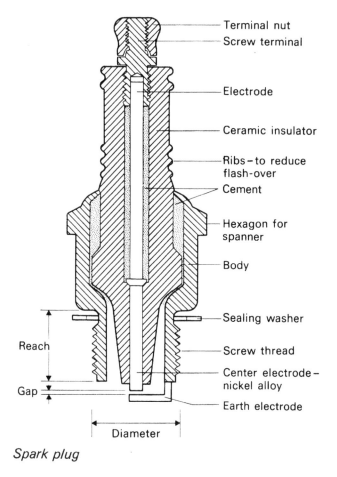

Spark plug

Figure 2.1 Cross-sectional view of spark plug. These come in a wide range of types and sizes. The ceramic part of the body usually carries a reference number for reordering.

list or the workshop manual should be used to check the exact code for the model. It is common for different specification or year models of vehicles to use different spark plugs.

Servicing spark plugs is limited to cleaning and gapping between replacements. If cleaning is needed, then a special machine is needed. Gapping the plug means setting the size of the **gap** between the fixed and the side electrode. Spark plug gaps are usually between 0.020 and 0.040 in (0.5 and 1 mm). New spark plugs are usually already gapped from the factory, but it is worth checking them before fitting. Plug condition is usually indicated by the engine analyzer test.

40 Motorcycle Engineering

Figure 2.2 Spark plug in a single-cylinder 50-cc engine – it's a good idea to carry a spare spark plug and socket tool to change it in case of failure.

Tech Note

Setting spark plug gaps takes skill and practice. Hold the feeler gauge on the flat surfaces between your first finger and thumb. The correct clearance can be felt as a faint touching drag when you move the feeler gauge through the gap. Close the gap by gently tapping the side electrode on a hard metal surface such as a bench top. Do not use a screwdriver or pliers to alter the gap. Before fitting new spark plugs always check the gap settings. When tightening up spark plugs always use a correctly set torque wrench. The taper-fitting spark plugs must not be overtightened. If the spark plug thread in the cylinder head is slightly rusty or, in an aluminum cylinder head, dry, apply a little light oil to the plug threads before fitting them.

Distributor

The inside of the distributor is divided into three parts. The lower part houses the mechanical components and the linkages, the middle part comprises the LT components, and the upper part is the HT section. The HT components are mainly the **distributor cap** and the **rotor arm**. HT electricity is delivered from the coil to the center of the cap.

Electricity flows from the cap to the rotor arm through a **brush** arrangement. The **rotor arm** is connected to the **distributor spindle** so that it goes around at the same speed as the **spindle**. As the rotor arm rotates, its free end aligns with **segments** in the cap, one at a time. The sequence is the firing order. As each segment is aligned, the current is passed from the rotor arm to the segment. The segments each have a **plug lead** attached to carry the electricity on its way to the spark plug.

The LT part of the distributor comprises the **contact breaker (cb) points** and the **capacitor**. The **cam ring,** which is formed on the outside of the distributor spindle, rotates at the same speed as the spindle. As the cam goes round, it opens and closes the cb points. It is this action which causes the current to flow in the HT circuit by **induction** in the coil. The **gap** of the cb points, and their position in relation to the spindle, affects the **ignition timing** and general efficiency. The cb points should be checked for condition and size of gap every 8000 km (5000 m). They should be replaced every 16,000 km (10,000 m).

The **capacitor** (also called a **condenser**) is fitted to give a good-quality spark by controlling the flow of electricity; this reduces **arcing** at the points and gives a longer life to the cb points.

The drive for the distributor is picked up from the camshaft by an angular gear called a **skew gear**. This skew gear is on the lower end of the distributor spindle. The spindle passes through the mechanical and the vacuum timing mechanisms. These mechanisms advance and retard the ignition timing to suit different engine speeds and load characteristics.

Tech Note

Does the distributor rotate at the same speed as the crankshaft or that of the camshaft?

The distributor rotates at the same speed as the camshaft, that is, half the speed of the crankshaft. Therefore, on a four-cylinder engine the cb points open four times for each revolution of the distributor spindle.

Figure 2.3 Harley Davidson coil pack. This Vee-twin engine has four spark plugs for reliability.

Ignition timing

The ignition system is designed so that the spark occurs in the combustion chamber a small number of degrees **before top dead center** (BTDC). Motorcycle manufacturers give specific figures for each of their different models. Typically, the static timing is 10 degrees BTDC; that is, when the engine is rotated by hand. When the engine is running – dynamic timing – the timing will advance to about 30 degrees BTDC.

Tech Note

You will find the ignition timing settings in the workshop manual or service data book. The static timing can easily be checked using a small low-wattage bulb (such as a side lamp bulb) in a holder with two wires attached. Attach one wire to the distributor terminal on the ignition coil – usually marked with a positive sign (+). Attach the other wire to a good chassis earth. Disconnect the HT king lead or remove the distributor cap so that the engine will not start. Switch on the ignition and turn the engine by hand. The bulb should light just as the timing mark comes into line. The timing can be adjusted by slackening the distributor clamp and moving the distributor gently. With the timing marks aligned, move the distributor until the bulb just lights.

Ignition and fuel 43

Figure 2.4 Stroboscopic timing light to enable checking of the timing with the engine running.

Dwell angle

This is the period for which the contact breaker points remain closed. When the cb points are closed, the magnetic field is building up in the ignition coil. When the points open, the spark is triggered at the plugs. The dwell angle is directly proportional to the number of cylinders and the cb points gap, and as such is used as an indicator of the engine condition. For four-cylinder engines it is typically 60 degrees.

Electronic ignition

From the 1970s to about the 2000 model year lots of different types of electronic ignition systems were tried and discarded. They were mainly based on the CDI system; they have now been almost completely replaced by digital systems. The two classes are:

1. Capacitor discharge ignition (CDI) – an early form of electronic ignition
2. Digital electronic ignition using an ECU – distributor-less ignition system (DIS)

First, we will look at some of the details of the CDI concept. The distributor has an electronic trigger device instead of the cb points. On the outside of the distributor is an amplifier unit. The switch, coil, and battery remain the same. The distributor may be incorporated in the alternator or end of the crankshaft. The reason for the move to CDI was to get a stronger spark at the plug for the higher compression ratios of the modern engines. That is about 40kV instead of 10 or 15kV on the older engines. The principle is for the LT supply to charge a capacitor to about 350V. When the system is triggered – and, of course, you cannot do that with cb points – the 350V goes to the coil and is converted to 40kV compared to 12V of a normal system being taken to 10kV.

Three main types of trigger devices are used instead of cb points:

- **Inductive type** – This uses a magnet attached to the distributor shaft to induce a small electrical current into a pick-up coil. The current produced is low voltage (typically 2V) alternating current (AC).
- **Hall effect** – This uses a small integrated circuit (IC) as a switch, which is turned on and off by the passing of a metal drum-shaped component. This works at low voltage (typically 5V).
- **Optical type** – This uses an infrared light-emitting diode (LED), which shines on a phototransistor. The light is turned on and off by a Maltese cross–shaped component usually referred to as a light chopper. Operating voltage is typically 9V.

As the electronic ignition system does not have cb points, there are fewer moving parts, so reliability is increased along with engine economy and performance. The only ignition servicing that is required is changing the spark plugs. On electronic ignition motorcycles, the spark plugs may only need changing every 60,000 km (40,000 m).

Other types of electronic ignition systems do not have a distributor and do not have a conventional coil. Also, the ignition lock and key have become a more complicated security device. The key incorporates a very small electronic device called a transponder; this transponder is a sort of electronic key that unlocks the ignition at the same time that it is being mechanically unlocked.

Distributor-less Ignition System

The DIS is usually part of an engine management system that comprises two parts: **ignition** and **fuel**. Looking at the ignition part the DIS includes:

- **Ignition IC** in the **ECU**
- **Sensors** for **crankshaft position, top dead center (TDC), knock** (pinking), **throttle position,** and **engine temperature**
- **Spark plugs**
- **Ignition coils** for each spark plug

Ignition and fuel 45

> **Tech Note**
>
> Sensors on different makes of systems may vary and have different values for different engines.

The ignition IC is **programmed** to trigger a spark according to engine load and road conditions – it is fully self-regulating and needs no servicing apart from spark plug replacement at set intervals. In the event of a fault, this is found using **electronic diagnostic equipment**.

FUEL SYSTEM

Figure 2.5 Single carburetor on 50-cc engine.

> **Key points**
>
> - Petrol must be handled with care. It is highly flammable, and the fumes can be a hazard to your health.
> - Petrol engines mainly use fuel injection, but some older motorcycles have carburetors.
> - Air filters and fuel filters need changing regularly.
> - The fuel system must be set up to supply the correct amount of fuel at the correct time.
> - Correct servicing is needed to both maintain fuel economy and reduce environmental pollution.

The fuel system supplies the motorcycle with the necessary amount of fuel for it to be able to do its work efficiently. The engine must receive the correct amount of fuel at the right time or else it will not run properly, if at all. Fuel is the food of the engine. Petrol is a hydrocarbon; it is made up of hydrogen and carbon atoms.

It is important to remember that burning fuel creates a hot exhaust gas, which, if the fuel system is not set up correctly, can cause illegal pollution.

> **SAFETY NOTE**
>
> Before working on a petrol supply system, you should be aware of the following hazards:
>
> - Petrol is highly flammable. Do not smoke or have any naked lights near petrol.
> - Petrol dries the skin and can cause skin disease. Always wear protective gloves and avoid direct contact.
> - If your overalls are doused in petrol at all, change them. There is an extreme fire risk even after the petrol has dried.
> - If draining a fuel tank, a sealed and electrically earthed draining appliance must be used.
> - Petrol must not be stored in the workshop.

Carburetor – This mixes the petrol and the air in the correct proportions for it to be burnt inside the combustion chamber.

Figure 2.6 Air cleaner – you should check and clean this regularly.

The simple carburetor

The basic principles of carburation are embodied in the simple carburetor. The simple carburetor was used on the very first cars and motorcycles and can be found on lawnmowers and other single-speed garden equipment. The petrol is in the **float chamber**, which is connected to the **discharge jet** through a tube that contains the main metering jet. When the engine is spun over with the throttle valve and the **choke butterfly** open, air is drawn into the **venturi**. As the air passes through the narrow section of the venturi, its speed increases. Increasing the air speed causes the air pressure to drop so that the air pressure in the venturi is below normal atmospheric pressure. The petrol in the float chamber is at atmospheric pressure; it now flows through the main metering jet and out of the discharge nozzle to mix with the air. As the petrol leaves the discharge nozzle, it breaks into very small droplets; this is referred to as **atomization**.

48 Motorcycle Engineering

Figure 2.7 Fuel tap on tank – turn the fuel off when the engine is not running to prevent flooding.

Figure 2.8 Secondary fuel tap – remember that you need both on for the engine to run.

Ignition and fuel 49

Figure 2.9 Locking fuel cap on top of petrol tank.

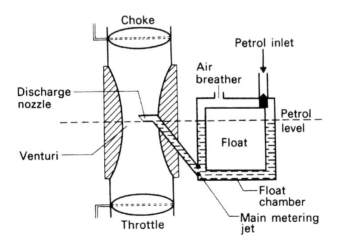

Simple carburetor

Figure 2.10 Simple carburetor.

NOMENCLATURE

Venturi is the name given to the narrowing of an air passage. In a carburetor this is sometimes referred to as the choke tube, or simply choke.

The **throttle flap** controls the flow of petrol and air through the carburetor; on a car this is operated by the throttle pedal. The **choke flap** controls the mixture strength by limiting the air flow into the carburetor. The choke is closed for cold starting to give a richer mixture.

The flow of petrol into the carburetor float chamber is controlled by the **needle valve** and float. As the **float chamber** fills up with petrol, the float rises. The rising float pushes the needle valve up against its seat and stops the flow of petrol.

AIR:PETROL RATIO

For the complete combustion of petrol and air, approximately 14 parts of air are needed to burn 1 part of petrol. However, when an engine is running under different conditions, the mixture strength needs to be altered. Table 2.1 shows typical mixture strengths for different operating conditions.

Adjusting the carburetor – You can usually adjust both the **mixture strength** and the **idling speed** of the carburetor. The workshop manual will give information on both of these settings.

FAQS: HOW ARE CARBURETOR ADJUSTMENTS CHECKED?

The idling speed on most carburetor engines is between 600 and 900 rpm. Generally, if it sounds nice, it is correct. Because of the exhaust emission regulations, you must set the mixture using an exhaust gas analyzer – cheap toolbox ones are available for the home mechanic on a tight budget.

Table 2.1 Air:fuel ratios

Operating condition	Air:fuel ratio
Cold starting	9:1
Slow running	13:1
Accelerating	11:1
Cruising	19:1

Ignition and fuel 51

Figure 2.11 Screaming Eagle performance air filter on Harley Davidson.

Air filter – The air filter assembly has three functions, namely:

1. To filter and clean the incoming air and prevent the entry of dust, grit, and other foreign bodies into the engine, which could seriously damage the engine.
2. To silence the air movement, thus making the engine less noisy.
3. To act as a flame trap, thus preventing a serious under-the-bonnet fire should the engine backfire.

Most vehicles are fitted with paper element air filters. These must be replaced at set mileage intervals – usually about 20,000 miles. Air filters must be replaced regularly even if they look clean; their micropore surface may still be blocked.

Tech Note

Fuel consumption in SI units is given in liters per hundred kilometers (liters/100 km). In imperial units it is in miles per gallon (mpg). For conversion purposes 9 liters/100 km approximately equals 30 mpg.

PETROL INJECTION

This section covers the petrol injection system used on most new motorcycles – variations are used for specialist applications.

All new motorcycles are fitted with petrol injection (PI). Petrol injection machines have a similar petrol tank, petrol pipes, and air filter. The carburetor is replaced by injectors and a high-pressure pump. PI systems are usually more efficient than carburetors in that they give more power, better fuel consumption, and more controllable emissions; they also need less maintenance. Let's have a look at the main components.

Fuel (petrol) pump

The petrol pump on a PI engine is likely to be electrically operated and submerged in the petrol tank. It is submerged to keep it cool and prevent the entry of air bubbles into the petrol pipes. The PI engine petrol pump raises the petrol pressure to about 1650 kPa (110 psi) and sends it to the injectors via a filter. The petrol pipes on a PI system are much heavier than those on a carburetor engine car, and the pipe couplings have screw threaded ends. The petrol pump is controlled electronically by the ECU.

ELECTRONIC CONTROL UNIT

The ECU is a sealed box containing a number of microprocessor integrated circuits – usually referred to as microchips – similar to those in a computer. The ECU controls both the idle speed and the mixture strength, so there is no need to adjust these settings. The ECU for the PI equipment also controls the ignition system.

Injectors

There may be one centrally mounted injector, called single-point injection, or one for each cylinder. The petrol under pressure is supplied to the injectors from the petrol pump. The ECU electronically controls the amount of fuel to be delivered by the injectors.

Injectors open and close about 50 times each second when the engine is running at full speed, delivering petrol on each occasion. Injectors can become blocked or worn. To maintain injectors in good working order, they should be removed and checked at regular service intervals. Their spray pattern is observed, and they may be cleaned or replaced as needed. Cleaning may be by ultrasonic vibrations or the use of a chemical cleaning agent. Injector cleaning chemicals are available to be added to the petrol in the tank.

Ignition and fuel 53

> **NOMENCLATURE**
>
> In this book we have used the term petrol injection and the abbreviation PI because it is a general term. You will come across electronic fuel injection (EFI), just injection (I or i), gasoline injection (GI), and a range of equivalent terms in different foreign languages.

Air filter – On a PI engine this is similar to the one used on carburetor engines in cars.

Air flow control

To control the flow of air into the engine, and hence the engine speed, a throttle body similar to that in a carburetor is used. This is connected by the accelerator cable to the accelerator (or throttle) pedal. The throttle butterfly is connected electrically to the ECU.

Figure 2.12 Lambda sensor to help control exhaust emissions.

Chapter 3

Lubrication and cooling

LUBRICATION SYSTEM

Lubrication and cooling fit together like eggs and bacon. The lubrication system cools the inner parts of the engine as well as lubricates them; the cooling system cools the larger parts of the block and head. The radiator and the oil cooler usually occupy juxtapositions at the front of the cold air intake grill.

> **Key points**
> - Lubrication is needed to reduce friction and wear.
> - The two main types of lubrication are full-film and boundary.
> - There are many different types of oil; it is important to choose the correct one.
> - Most engines use a pressurized oil supply using an oil pump, sump, filter, and pressure relief valve.
> - Engines and other units with rotating parts need oil seals to keep the oil in and the dirt out.
> - Two-stroke engines use total loss lubrication; this is only suitable for small motorcycles, as the burnt oil causes air pollution.

Lubrication is needed to keep mating bearing surfaces apart; this reduces friction and wear. The lubricant also acts as a coolant, taking heat away from the bearing surfaces to maintain a constant running temperature. The liquid cooling system runs at about 85 °C; the lubrication system runs at a slightly higher temperature: 90° to 120 °C are typical figures for road use; for racing engines, this may exceed 200 °Celsius. The lubricant also picks up small particles of metal and carbon from the components that it passes over and deposits them in the oil filter. The particles that are too small to be filtered out make the oil a dirty black color.

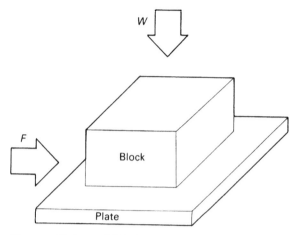

The coefficient of friction

Figure 3.1 The coefficient of friction.

Tech Note

Oil temperature gauges are typically red-lined at about 150 °C.

Friction

Friction is the resistance of one surface to sliding over another. The amount of friction is referred to as the coefficient of friction and is indicated by the Greek letter mu that has the symbol μ.

To calculate the coefficient of friction, the formula used is:

$\mu = F / W$

where F is the force that is required to slide the object over the mating surface, and W is the weight of the object. Both F and W are given in the same units, usually newtons (N). The value of μ is always less than unity; that is between 0 and 1.

Lubrication keeps the mating surfaces apart so that they can slide easily over each other. The simplest form of lubricant is water – think of how slippery a floor is when it is wet. Motor vehicles use a variety of oils and greases in a range of different ways—the following sections look at some of them.

NOMENCLATURE

μ almost always has a value of less than unity (in other words 1), as it is usually easier to push something along than to lift it.

Lubrication and cooling 57

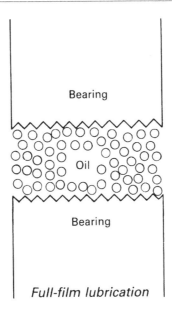

Figure 3.2 Full-film lubrication.

Types of lubrication

Lubricants are generally used in the form of oil, which is liquid, or grease, which is semi-solid. Oil in the engine is fed to the bearings under pressure; the film of the oil keeps the bearing surfaces apart; this is known as full-film lubrication. Grease, which is used for steering and suspension joints and wheel bearings, does not keep the bearing surfaces fully separated; this type of lubrication is known as boundary lubrication.

Viscosity

Oils are classified according to their viscosity. **Viscosity is the resistance of an oil to flow.** It is calculated by timing how long it takes a fixed quantity of oil to flow through a specific-diameter aperture at a preset temperature. The aperture is a hole in a piece of metal that looks like a washer. The time is in seconds. Because the timing is in this specific way, the seconds are called Redwood seconds. The longer the amount of time taken, the higher the oil's viscosity is said to be and the higher the viscosity number it is given. In common language it is said to be thicker. Basic oils are:

5 SAE	Cycle oil
30 SAE	Straight engine oil (not commonly used)
90 SAE	Gear/axle oil

SAE is the abbreviation for the **Society of Automotive Engineers**; this is an American-based international organization that sets standards for many areas of automotive engineering.

When the temperature of oil is raised, its viscosity usually decreases; that is to say when it gets hotter, it gets thinner too. Therefore, most modern oils for vehicle engines are multigrade. They have two viscosity ratings: one for when they are cold and one for when they are hot. A typical oil is 15/40 SAE. This oil is rated as a thin 15 SAE when it is cold to give easy cold starting; it is rated as a thick 40 SAE when it is hot to give good protection to the engine, such as on the motorway. Other popular grades are 10/40 SAE and 5/30 SAE.

Types of oil

Oil can be classified in a number of different ways; it is important to choose the correct oil for the vehicle. Let's look at some of the common types:

- Mineral oils are those pumped from the ground, and vegetable oils are those, such as Castrol R oil, that are made from vegetable products. Castrol R oil is used for a small number of specialist applications such as racing motorcycle engines. Most motorcycles use mineral oils. Mineral and vegetable oils cannot be mixed.
- Synthetic oils are chemically engineered oils; they are specially prepared mineral oils. Synthetic oils are very expensive. A cheaper alternative is semi-synthetic oil.
- Oil for diesel engines and ones with turbochargers are specially classified to withstand the very high temperatures and pressures of these engines – look for the marking on the container. Synthetic and other oils may be used for these special purposes.
- Extreme pressure (EP) oils are used in gearboxes; usually their viscosity rating is 80 or 90 SAE. They maintain a film of oil under a very heavy load between two gear teeth.
- Hypoy (or hypoid) oil is used for a special shape of gear teeth in shaft drive and tricycle axles. It is usually of 90 SAE or 140 SAE viscosity.
- Automatic transmission fluid (ATQ or similar) is an oil that is also a hydraulic fluid. It is used in power steering as well as the automatic gearbox.
- Three-in-one oil, or cycle oil, is 5 SAE and is used because it cleans and prevents rusting as well as lubricating. On motorcycles it is used for control cables, hinges, and electrical components.

Tech Note

Because of the exacting nature of lubrication in motorcycle engines, it is essential to use the right oil irrespective of cost – a race engine rebuild may cost more than the price of a small family car. Many of the specialist race lubrication suppliers and manufacturers sponsor racing teams because they enjoy the sport – not just to sell more oil.

Engine lubrication system

The most popular type of engine lubrication is the wet sump type. This is so called because the sump is kept wet by the engine oil, which uses the sump as a supply reservoir.

In operation the oil is drawn from the sump by the oil pump. To prevent foreign bodies from being drawn into the pump, it passes through a gauze strainer at the lower end of the pick-up pipe. From the pump the oil is passed through a paper filter and then through a drilling in the cylinder block to the main oil gallery. The oil pressure in the filter and main gallery is about 400 kPa (4 bar or 60 psi). To ensure that the design pressure is not exceeded, and thus prevent damage to the oil seals, an oil pressure relief valve is fitted in the main gallery. The oil, which is still at a pressure of about 400 kPa, goes through drillings in the block to the main bearings. The oil then passes through drillings in the crankshaft to the big end bearings. At one end of the engine there is a long vertical drilling to take the oil supply to the camshaft and valve gear.

> **Tech Note**
>
> Oil pressure gauges typically read in psi, which may also be written lb/in^2. During normal usage, the oil pressure should remain constant within a given range – surges or drops outside of this range indicate possible faults.

Gear-type oil pump

The function of the oil pump is to draw oil from the sump and send it under pressure to the filter and on to the main gallery from where it is distributed to other parts of the engine. The gear-type oil pump is used in a large number of vehicles. It is driven by the camshaft, usually by a skew gear mechanism. The skew gear drives only one of the two toothed gear wheels in the pump; the other follows the driven gear. Oil enters the pump through the inlet from the pick-up pipe. Oil is carried round the pump in the spaces between the gear teeth; it is then forced out of the pump into the filter. The meshing of the gear teeth on their inside prevents the oil returning to the sump. If the teeth wear so that the clearance between the gear teeth and the casing exceeds 0.1 mm (0.004 in), the oil pressure will be reduced and the engine might become noisy, especially when it starts from cold when a big-end bearing rattle might be heard.

Oil pressure relief valve

To prevent the engine oil from exceeding a pre-set figure, an oil pressure relief valve is located in the main oil gallery. Excess pressure may be generated if there a blockage in the gallery or the engine runs at high speed for a period. The oil pressure relief valve takes the form of a spring-loaded plunger, which

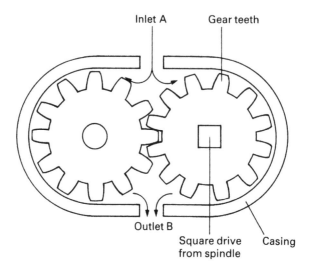

Figure 3.3 Gear-type oil pump.

is forced away from its seat when the pre-set pressure is reached. Excess oil from the gallery passes through the valve and returns directly to the sump. Removing oil from the main gallery reduces the oil pressure so that the pressure is reduced. When the pressure is lowered the valve is closed by the force of the spring, the oil can no longer return to the sump.

Oil pressure warning light

The purpose of the oil pressure warning light on the dashboard is to warn the driver of low oil pressure. The dashboard light is activated by the pressure-sensitive switch that is screwed into the main oil gallery. The switch makes the light glow when the oil pressure drops below about 30 kPa (5 psi). The light will also come on when the engine is switched on but is not running. Higher-rated switches are available for tuned engines – typically 100 kPa (15 psi).

Oil filter

It is necessary to keep the engine oil clean and free from particles of metal and carbon that could damage the inside faces of the oil pump or bearing surfaces. This is the job that the oil filter performs. All the oil passes through the filter before it goes to the bearings. The oil filter element is made from a special porous paper; this allows the oil to pass through it but hold back the foreign bodies. The oil filter element is housed in a metal canister that works as a sediment trap for the particles that the filter element has prevented from flowing into the main gallery.

Lubrication and cooling 61

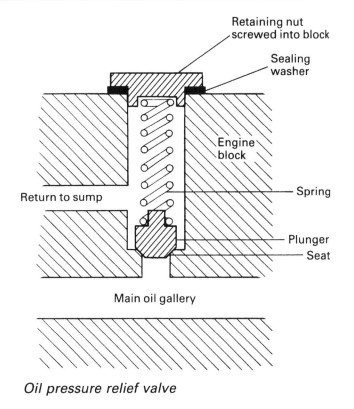

Oil pressure relief valve

Figure 3.4 Oil pressure relief valve.

NOMENCLATURE

When the moving parts of the engine rub against each other, they wear down. That is, tiny particles of metal are scraped off or dislodged into the sump or other parts of the engine. Also, the heat and combustion inside the engine create particles of carbon or soot. These particles are referred to as foreign bodies. In filters, the heavier foreign bodies drop off the filter walls and fall to the bottom of the canister, or filter housing; this layer of trapped material is referred to as sediment.

Servicing the engine lubrication system

Servicing the engine lubrication system entails changing the engine oil and replacing the oil filter. This is done at the intervals set by the vehicle manufacturer. Typically, this service interval is 16,000 km (10,000 mi). When carrying out lubrication system servicing, the following points should be noted:

Figure 3.5 Sight glass to check oil level.

- The rubber seal on the oil filter must always be replaced.
- The sump plug's sealing washer must always be checked and replaced if necessary.
- Always check the oil level with the vehicle on a level surface.
- After the engine has run, wait for about two minutes before rechecking the oil level and adding more oil.
- Never overfill or underfill the engine; keep the oil level between the minimum (MIN) and the maximum (MAX) marks.

SAFETY NOTE

Beware of hot engine oil. Never change the oil straight after the vehicle has completed a long or fast journey, as the oil temperature can be over 150 °C; therefore, there is a risk of scalding when removing the sump plug or filter. Always use barrier cream and protective rubber gloves when dealing with dirty engine oil to reduce the risk of contracting dermatitis.

Oil seals

To prevent oil leaks from rotating components, an oil seal is needed. There are three different types of oil seals:

- Felt packing
- Scroll seals
- Lip seals

Felt packing, as its name implies, is soft felt packed into a cavity. The rotating shaft turns in gentle contact with the felt. The felt becomes soaked in oil; this reduces the friction against the shaft. The oil also expands the

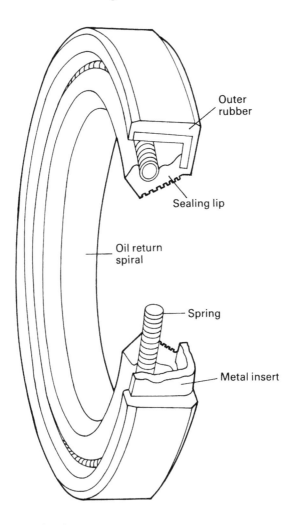

Figure 3.6 Lip-type oil seal.

felt so that the seal remains tight on the shaft and provides an effective oil seal.

The scroll is a groove cut in the rotating shaft that acts like an Archimedean screw. That is, as the shaft rotates, the screw draws the oil away from the open end of the shaft and returns it to the sump.

The lip seal forms a hard, spring-loaded knife edge against the rotating shaft. The pressure exerted by this knife edge is very high and prevents any oil from flowing past the seal under even the most arduous conditions. The lip seal is retained in the oil seal housing by the interference fit of its outer rubber layer.

Two-stroke petrol engine lubrication

Lubrication of the two-stroke engine is by a petrol–oil mixture. The lubricating oil is added to the mixture of petrol and air so that it lubricates the crankshaft and big-end bearings when it is in the crankcase while on its way to the combustion chamber.

The burned lubricating oil is emitted from the exhaust in the form of blue smoke. Because of the effect this has on the environment, two-stroke engines are only allowed in small motorcycles.

Low-performance two-stroke engines have their supply of lubricating oil added to the petrol when the tank is topped up. The oil lubricates the main and big-end bearings when the mixture is passing through the crankcase. The rider must add a quantity of oil to the petrol each time the motorcycle is refueled. The amount of oil will depend on the ratio of petrol to oil that is required by the engine. A typical ratio is 40:1 (40 parts of petrol to 1 part of oil). In imperial units this works out to 1 gallon of petrol being mixed with one-fifth of a pint of oil. Often the petrol filler cap is designed to incorporate a measure of the correct amount of oil to be added to 1 gallon of petrol. In SI units, this will be one-eighth of a liter of oil to 5 liters of petrol. The oil used is special two-stroke oil that is made to mix easily with the petrol. However, it is normal practice to rock the motorcycle from side to side after the petrol and oil have been added to ensure that it is a good mixture.

Total loss system

High-performance two-stroke motorcycles have a separate oil tank, usually located under the rider's seat. Oil is pumped from this by a crankshaft-driven pump so that it is mixed with the petrol at the carburetor. This ensures a constant mixture of petrol and oil in the correct proportions. It saves measuring out quantities of oil on the service station forecourt. Sometimes an adjustable regulator is fitted so that the petrol–oil mixture can be adjusted to suit different engine running conditions.

THE COOLING SYSTEM

> ### Key points
>
> - The purpose of the cooling system is to keep the engine at a constant temperature while preventing the overheating of any of the individual components.
> - Typically, petrol engines run at between 80° and 90 °C (180° and 190° F). This is the best temperature to get the best fuel consumption and the least pollution.
> - There are two main types of cooling systems: liquid (water) cooling and air cooling.

Liquid (water) cooling system

> **SAFETY NOTE**
>
> Coolant in an engine is likely to be scalding hot and under high pressure. Do not touch cooling system components or remove filler caps when hot.

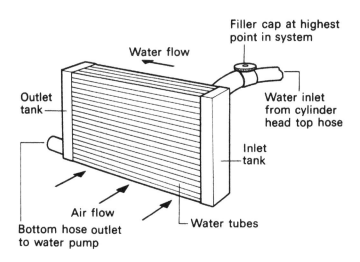

Cross-flow radiator

Figure 3.7 Cross-flow radiator.

The liquid cooling system works by using coolant, the name for water mixed with other chemicals, to remove the heat from the cylinder block and pass it to the radiator so that it is cooled down. That is, the coolant circulates through the engine, where it gets hot, then through hoses to the radiator, where it cools down again, and finally, back through another hose to the engine to go through the process again. When the petrol is burning in the combustion chamber, the temperature gets very hot, about 2000 °C (3500 °F). The cooling system therefore needs to work very hard to keep the temperature of the components at about 85 °C (185 °F). Most engines run at between about 80° and 90 °C (180° and 190 °F). The temperature of the engine is kept between 80° and 90 °C (180° and 190 °F) because this is the most efficient temperature; that is, it will use the least fuel and produce the least pollution.

Coolant has a natural tendency to circulate when heated up in the cooling system. The hot liquid rises, and the cooling liquid falls. This is called thermo-siphon. As the coolant is heated up in the cylinder block water jackets, it rises up; it then passes through the thermostat into the top hose to the radiator header tank. The water then falls through the radiator core into the bottom tank. As the coolant falls inside the radiator core, it is cooled by the incoming air, which passes around the outside of the radiator fins. The incoming air is from the front of the vehicle and may be forced along by the fan. The weight of the water in the radiator forces the coolant through the bottom hose back into the engine. The coolant (water) pump helps the coolant circulate more quickly into the water jackets. For the coolant to be able to circulate, the water level must be kept above the top hose connection so that it can maintain a flow into the radiator

The coolant

The liquid used in a cooling system is usually called coolant, because it cools the engine. The coolant is a mixture of water, antifreeze, and a chemical that inhibits corrosion to the metal parts inside the engine. As you probably know, water boils at 100 °C (212 °F) and freezes at 0 °C (32 °F). This means that in winter water could freeze and damage the engine. Typically the coolant mixture boils at about 110 °C (230 °F) and freezes at about −18 °C (0 °F).

Tech Note

Antifreeze is a chemical called ethylene glycol.

Coolant can be bought ready-mixed. This is usually advised for specialist engines, or for most vehicles antifreeze, which contains the anticorrosion chemicals, can be mixed with water. For British winters, it is normal to mix a

33% antifreeze solution, that is, one-third antifreeze and two-thirds water – the solution must be measured accurately.

The strength of the coolant can be checked with a special hydrometer called an antifreeze tester; usually this test must be carried out when the coolant is cold.

Key points

When mixing anti\freeze:

- Always use clean water to prevent damage to the engine
- Check the instructions on the label
- Mix the antifreeze and water before putting it in the engine

---

Tech Note

Coolant is one of the life bloods of an engine (the other is the oil). So always use the most appropriate coolant for your engine. Look for coolant that contains a wetting agent as well as a corrosion inhibitor and antifreeze. Aluminum engines are prone to internal corrosion if left partially drained.

Tech Note

Antifreeze loses it properties after about 2 years, so change it every 2 years or at the major services as recommended by the vehicle manufacturer.

NOTE FOR FUTURE RACING CHAMPIONS

Coolant temperatures will vary with race conditions; however, whether you win or not, you should keep the champagne in the fridge between 4° and 6 °C (39° and 42 °F).

Coolant (water) pump

To circulate the coolant quickly, a pump is fitted; this is called a coolant or water pump. The coolant pump is fitted to the front of the engine. The bottom hose from the radiator is fitted to the coolant pump inlet so that it is supplied with cooled coolant; the pump outlet is connected directly to the water jacket of the cylinder block. The coolant pump is typically driven off the same shaft as the oil pump.

Fan

The fan is used to draw in air through the grill or ducting and pass it over the outside of the **radiator fins** so that it cools the coolant inside the radiator core.

The fan is usually electrically driven. It is operated by an electric motor that is only switched on when the engine gets hot. There is a temperature-sensitive switch mounted on the radiator, so that when the radiator reaches a preset temperature, the fan is switched on; when it cools down, it is switched off.

Radiator

The radiator is made up of the **header tank**, the **bottom tank**, and the **core** between them. The flattened tube type of construction is used on most vehicles; heavy goods vehicles tend to use the round tube variety; the honeycomb design is only used on a few luxury or high-performance cars. The radiator fins are designed to increase the surface area of the cooling zone. That is, the fins dissipate, or spread out, the heat more efficiently so that it is transferred to the air quickly.

The radiator may be either **vertical** in design – with top and bottom tanks stacked with the core – or have a **cross-flow** design. The cross-flow radiator has its tanks mounted at each end. This means that the coolant flows from one side of the radiator to the other, rather than from the top to the bottom; this allows the radiator shape to be changed to short and wide rather than tall.

Thermostat

The thermostat is fitted in the outlet from the engine; it is a sort of tap to control the coolant flow between the engine and the radiator. The thermostat is usually in a special connector – one end of the connector holds the thermostat against the inside of the cylinder head, and the other end is attached to the top hose.

When the thermostat is closed, the coolant cannot flow; when the thermostat is open, it can flow freely. The thermostat allows a quick warm-up period by remaining closed until the engine has reached its required temperature. It keeps the engine at a constant temperature by opening and closing as the engine becomes hot or cools down.

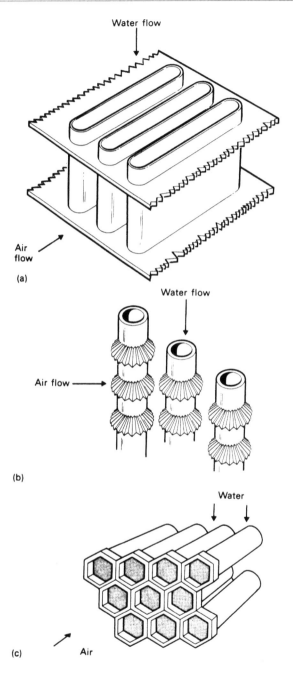

Radiator cores. (a) Flattened tube; (b) round tube; (c) honeycomb

Figure 3.8 Types of tubes found in radiators.

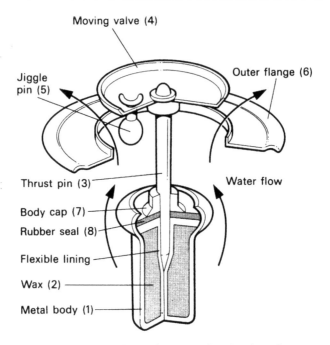

Wax thermostat (open). The valve is closed by an external spring (not shown).

Figure 3.9 Wax-stat thermostat.

Tech Note

A wax-stat is a **wax**-operated thermo**stat**.

The metal body or capsule is filled with wax, which expands its volume very rapidly at the temperature at which its designers want it to open, which is usually around 80 °C (180 °F). So when the wax reaches the design temperature, it rapidly expands, forcing the thrust pin out of the capsule against the pressure of the return spring (not shown in Figure 3.9). The thrust pin lifts the moving valve to allow coolant to flow from the engine to the radiator through the top hose. When the wax cools, it contracts; the return spring closes the valve and returns the thrust pin into the capsule. The jiggle pin is fitted to prevent the formation of a vacuum in the engine water jacket by allowing small amounts of coolant to flow even when the thermostat is closed.

Tech Note

You'll find the opening temperature of the thermostat stamped on either the rim or the capsule as a two-digit number.

Radiator pressure cap

SAFETY NOTE

Never remove the radiator cap when the engine is HOT or RUNNING.

At normal atmospheric pressure, water boils at 100 °C (212 °F). At high altitudes, the boiling temperature is reduced. This also applies to the coolant mixture, though the temperatures may be slightly different. To prevent the engine from boiling and overheating, a radiator pressure cap is fitted. The force of the spring in the cap ensures that the coolant is kept under pressure. The higher the pressure, the higher the boiling point of the coolant in the system. To prevent the buildup of a vacuum inside the radiator, a vacuum valve and spring are fitted. The vacuum valve prevents the radiator from imploding, or collapsing inwards, when the coolant temperature is reduced, and the coolant pressure therefore decreases.

Key point

Radiator caps are made to operate at different pressures. The design pressure is usually written on the outside of the cap – this may be in pounds per square inch (psi) or bars.

The actual retaining cap is secured to the radiator neck with two curved sections. The radiator cap is fitted to the radiator in a similar way as the lid on a jam jar. The pressure seal is held in place against the top of the radiator

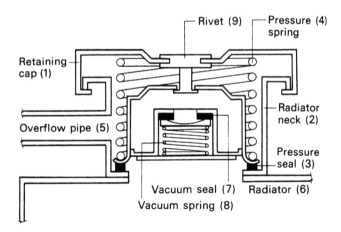

Radiator pressure cap

Figure 3.10 Radiator pressure cap.

by the pressure spring. Only when the coolant pressure exceeds the preset figure – which is stamped on the radiator cap – is the seal lifted against spring pressure. The coolant is then released through the overflow pipe. When the pressure of the coolant in the radiator is released by the coolant running off, the spring will press the seal back onto the top of the radiator. When the radiator cools, its coolant pressure decreases. Air pressure from outside the radiator forces the vacuum seal against the force of the vacuum spring to allow air to pass over the seal into the radiator.

Failure of the radiator pressure cap causes overheating and boiling over. The operating pressure of the cap can be tested with a special pressure gauge, and the condition of the seals can be inspected visually.

Sealed cooling system

To prevent the loss of coolant, and hence the need for topping up the radiator, most vehicles use sealed cooling systems. An **overflow tank** is fitted to the side of the radiator – sometimes this can be a little way from the radiator on the fairing panel. A tube from the radiator neck is connected to the overflow tank. Therefore, any coolant that is allowed past the radiator cap will go into the overflow tank. The tube is arranged so

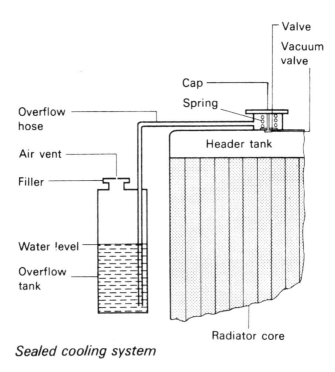

Figure 3.11 Sealed cooling system.

Lubrication and cooling 73

that its end is always below the coolant level (see Figure 3.11); therefore, when the radiator cools down and the coolant contracts, the coolant in the overflow tank is drawn back into the radiator to fill the available space. The overflow tank must be kept partially full to ensure that coolant covers the bottom of the tube.

Hoses

Rubber hoses are used to connect the engine to the radiator so that there is some flexibility between the two components. The engine is free to move slightly on its rubber mountings, but the radiator is rigidly mounted to the vehicle's body/chassis. The flexing of the hoses causes them to deteriorate. They usually crack, and if not replaced they will break or puncture, causing the total loss of coolant. The hoses are held in place with clips – the clips usually have a screw mechanism to tighten them up.

> ### Key points
> Hose clip drivers are available with hexagon ends. These are much safer to use and less likely to slip and cause damage to the radiator than a screwdriver.

When you are fitting a new hose, always check that the surfaces are clean and, if possible, use a new clip.

Figure 3.12 Radiator on Ducati Multistrada.

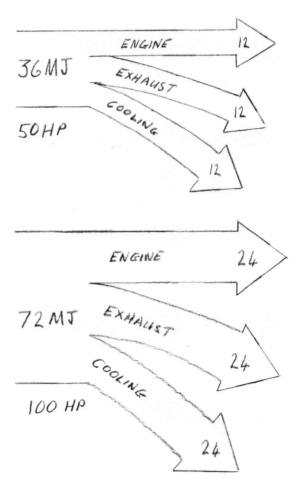

Figure 3.13 Sankey diagram. About a third of the heat from burning petrol goes to the cooling system. A 50-HP engine produces enough heat for a pair of semi-detached houses. This is the limit for air cooling. A 600-cc 100-HP engine produces enough heat for a small block of flats and therefore needs a very efficient cooling system.

Air cooling

Air cooling is used on older motorcycles. Air cooling has the advantages of not using a liquid coolant (water) and using fewer moving parts. Having no water, it cannot freeze or leak. However, air-cooled engines tend to be noisier than liquid-cooled ones.

Chapter 4

Health, safety, security, and the environment

Health and safety and the environment are controlled by a number of acts of Parliament and subsequent regulations and statutory instruments. These may have regional variations or specifics relating to the motorcycle and associated industries. These topics are also affected by other laws – for instance, those relating to the countryside. Perhaps the most important aspect of health and safety laws is that they are vicarious. That is, if you are a manager or supervisor in the motorcycle industry, you may be prosecuted for any injury or death caused by one of the technicians or other staff actions, as well as the staff involved being prosecuted. Breaking health and safety or environmental laws may result in custodial sentences as well as fines and damages to the injured parties. Meeting the requirements of the health and safety and environmental laws and regulations is the responsibility of everybody – ignorance of the law is not an excuse, so you need to take care.

PERSONAL HEALTH AND SAFETY PROCEDURES

Skin care (personal hygiene) systems. All employees should be aware of the importance of personal hygiene and should follow correct procedures to clean and protect their skin in order to avoid irritants causing skin infections and dermatitis. All personnel should use a suitable barrier cream before starting work and again when recommencing work after a break. Waterless hand cleaners are available that will remove heavy dirt on skin prior to thorough washing. When the skin has been washed, after-work restorative creams will help restore the skin's natural moisture.

Many paints, refinishing chemicals, and motorcycle repair shop materials will cause irritation upon contact with the skin and must be removed promptly with a suitable cleansing material. Paint solvents may cause dermatitis, particularly where skin has been in contact with peroxide hardeners or acid catalysts: these have a drying effect that removes the natural oils in the skin. Specialist products are available for the paint shop that will remove these types of materials from the skin quickly, safely, and effectively.

Hand protection. Motorcycle technicians are constantly handling substances that are harmful to health. The harmful effect of liquids, chemicals, and materials on the hands can be prevented in many cases by wearing the correct type of gloves. To comply with Control of Substances Hazardous to Health (COSHH) Regulations, vinyl disposable gloves must be used when refinishing paintwork to protect skin from toxic substances. Other specialist gloves available are rubber and polyvinyl chloride (PVC) gloves for protection against solvents, oil, and acids; leather gloves for hard-wearing and general repair work in the workshop; and welding gauntlets, which are made from specially treated leather and are longer than normal gloves to give adequate protection to the welder's forearms.

Protective clothing is worn to protect the worker and his or her clothes from coming into contact with dirt, extremes of temperature, falling objects, and chemical substances. The most common form of protective clothing is the overall, a one-piece boiler suit made from good-quality cotton, preferably flameproof. Nylon and other synthetic materials tend to be highly flammable and therefore pose a hazard in the vicinity of open flames. Worn and torn materials should be avoided, as they can catch in moving machinery. Where it is necessary to protect the skin, closely fitted sleeves should be worn down to the wrist with the cuffs fastened. All overall buttons must be kept fastened, and any loose items such as ties and scarves should not be worn. The coveralls must withstand continuous exposure to a variety of chemicals. They can be of the one-piece variety or can have separate disposable hoods.

Eye protection is required when there is a possibility of eye injury from flying particles when using a grinder, disc sander, power drill, or pneumatic chisel. Many employers are now requiring all employees to wear some form of safety glasses when they are in any workshop area. There is always the possibility of flying objects, dust particles, or splashing liquids entering the eyes. Not only is this painful but it can, in extreme cases, cause loss of sight. Eyes are irreplaceable; therefore, it is advisable to wear safety goggles, glasses, or face shields in all working areas. The following types of eye protection are available:

- **Lightweight safety spectacles** with adjustable arms and with side shields for extra protection. There is a choice of impact grades for the lenses.
- **General-purpose safety goggles** with a molded PVC frame that is resistant to oils, chemicals, and water. These have either a clear acetate or a polycarbonate lens with BS impact grades 1 and 2.
- **Face shields** with an adjustable head harness and deep polycarbonate brow guard with replaceable swivel-up clear or antiglare polycarbonate visor BS grade 1, which gives protection against sparks, molten metal, and chemicals.

- **Welding helmet or welding goggles** with appropriate shaded lens to BS regulations. These must be worn at all times when welding. They will protect the eyes and face from flying molten particles of steel when gas welding and brazing, and from the harmful light rays generated by the arc when MIG/MAG, TIG, or MMA welding.

Foot protection. Safety footwear is essential in the workshop environment. Boots or shoes with steel toecaps will protect the toes from falling objects. Rubber boots will provide protection from acids or in wet conditions. Never wear defective footwear, as this becomes a hazard in any workshop environment.

Ear protection. The **Noise at Work Regulations 2005** defines three action levels for exposure to noise at work:

- A daily personal exposure of up to 80 dB. Where exposure exceeds this level, suitable hearing protection must be provided on request.

Figure 4.1 Lightweight safety glasses.

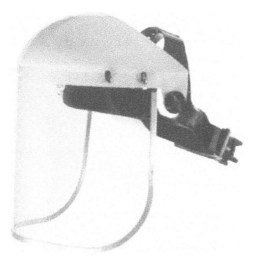

Figure 4.2 Face protection mask.

Figure 4.3 Automatic welding mask.

- A daily personal exposure of up to 85 dB. Above this second level of provision, hearing protection is mandatory.
- A peak sound pressure of 87 dB.

Tech Note

Noise is measured in decibels (dB) – this is measured on a scale based on logarithms. That is to say that increases do not follow the normal arithmetic scale in terms of increase in noise. An increase of 3 dB, from, say, 84 dB to 87 dB, will give a doubling of the noise heard. So, cutting the reading by 3 dB will reduce the noise heard by half.

Motorcycles are renowned for their noise. Continued exposure to this noise without ear protection is reckless. Use ear protection whenever it is needed.

Fire precautions

There is a detailed requirement for the recording of assessments, the provision of training, and the requirement that means of fighting fire, detecting fire, and giving warning in case of fire be maintained in good working order. The precautions vary with the size of the establishment, but in all cases they must be adequate and indicate appropriate fire exits.

What is fire? Fire is a chemical reaction called combustion (usually oxidation resulting in the release of heat and light). To initiate and maintain this chemical reaction, or in other words for an outbreak of fire to occur and continue, the following elements are essential:

Fuel: A combination substance: solid, liquid, or gas.

Oxygen: Usually air, which contains 21% oxygen.

Heat: The attainment of a certain temperature – once a fire has started, it normally maintains its own heat supply.

Methods of extinction of fire. Because three ingredients are necessary for fire to occur, it follows logically that if one or more of these ingredients are removed, the fire will be extinguished. Basically three methods are employed to extinguish a fire: removal of heat (cooling), removal of fuel (starving), and removal or limitation of oxygen (blanketing or smothering).

Removal of heat. If the rate of heat generation is less than the rate of dissipation, combustion cannot continue. For example, if cooling water can absorb heat to a point where more heat is being absorbed than generated, the fire will go out.

Removal of fuel. This is not a method that can be applied to fire extinguishers. The subdividing of risks can starve a fire, prevent large losses, and enable portable extinguishers to retain control; for example, part of a building may be demolished to provide a fire stop. The following advice can contribute to a company's fire protection program:

- What can cause fire in this location, and how can it be prevented?
- If a fire starts, regardless of cause, can it spread?
- If so, where to?
- Can anything be divided or moved to prevent such spread?

Removal or limitation of oxygen. It is not necessary to prevent the contact of oxygen with the heated fuel to achieve extinguishment. Where most flammable liquids are concerned, reducing the oxygen in the air from 21% to 15% or less will extinguish the fire. Combustion becomes impossible even though a considerable proportion of oxygen remains in the atmosphere. This rule applies to most solid fuels, although the degree to which oxygen content must be reduced may vary. Where solid materials are involved, they may continue to burn or smolder until the oxygen in the air is reduced to 6%. There are also substances that carry within their own structures sufficient oxygen to sustain combustion.

Fire risks in the workshop. Fire risks in the repair shop cover all classes of fire: class A is paper, wood, and cloth; class B is flammable liquids such as oils, spirits, alcohols, solvents, and grease; class C is flammable gases such as acetylene, propane, and butane, as well as electrical risks. It is essential that fire is detected and extinguished in the early stages. Workshop staff must know the risks involved and should be aware of the procedures necessary to combat fire. Workshop personnel should be

80 Motorcycle Engineering

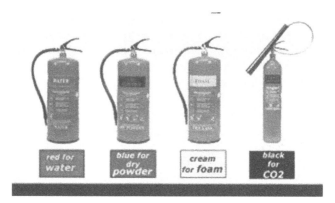

Figure 4.4 Fire extinguishers.

aware of the various classes of fire and how they relate to common workshop practice.

Class A fires: wood, paper and cloth. These fires often start by carelessness – throwing hot or burning objects in the waste bin. Always use waste bins with lids, which will prevent the spread of fires.

Class B fires: flammable liquids. Flammable liquids are the stock materials used in the trade for all motorcycle refinishing processes: gun cleaner to clear-finish coats and cellulose for the more modern finishes can all burn and produce acrid smoke.

Class C fires: gases. Lots of aerosol products contain flammable gases. These can be ignited by sunlight on hot days.

> **SAFETY NOTE**
>
> As motorcycle fuel systems are usually gravity fed from the fuel tank, leakages in the supply line can quickly cover the engine in petrol, and if the filler cap is left open, petrol fumes will quickly fill the surrounding area. It only takes a small spark, perhaps from disconnecting the battery, to ignite a major fire.

Electrical hazards. Electricity is not itself a class of fire. It is, however, a potential source of ignition for all of the fire classes mentioned earlier. The Electricity at Work Regulations cover the care of cables, plugs, and wiring. In addition, in the workshop, the use of welding and cutting equipment produces sparks that can, in the absence of good housekeeping, start a big fire. Training in how to use firefighting equipment can stop a fire in its early stages. Another hazard is the electrical energy present in all large batteries. A short-circuit across the terminals of a battery can produce sufficient energy to form a weld and in turn heating, a prime source of ignition. For instance, when tackling a motorcycle fire, a fireman will always try

Health, safety, security 81

to disconnect the battery; otherwise, any attempt to extinguish a fire can result in the reignition of flammable vapors.

Paste filler is a further possible source of ignition to be aware of. The result of mixing in the wrong proportions can give rise to an exothermic (heat-releasing) reaction; in extreme cases, the mix can ignite.

General precautions to reduce fire risk

- Good housekeeping means putting rubbish away rather than letting it accumulate.
- Read the manufacturer's material safety data sheets so that the dangers of flammable liquids are known.
- Only take from the stores sufficient flammable material for the job at hand.
- Materials left over from a specific job should be put back into a labeled container so that not only you but anyone (and this may be fire personnel) can tell what the potential risk may be.
- Be extremely careful when working close to plastic components.
- Petrol tanks are a potential hazard: supposedly empty tanks may be full of vapor. To give some idea of the potential problem, consider 5 liters (1 gallon) of petrol: it will evaporate into 1 m^3 (35 $feet^3$) of neat vapor, which will mix with air to form 14 m^3 (500 $feet^3$) of flammable vapor. Thus, the average petrol tank needs only a small amount of petrol to create a tank full of vapor.

The keys to fire safety are:

- Take care.
- Think.
- Train staff in the correct procedures before things go wrong.
- Ensure that these procedures are written down, understood, and followed by all personnel within the workshop.

Tech Note

Any carbon-based material will burn in air if at the temperature needed for combustion – be aware of this with dust in the factory or workshop.

Types of portable fire extinguishers

Water is the most widely used extinguisher agent. With portable extinguishers, a limited quantity of water can be expelled under pressure and its direction controlled by a nozzle.

There are basically two types of water extinguishers. The gas (CO_2) cartridge–operated extinguisher, when pierced by a plunger, pressurizes the body of the extinguisher, thus expelling the water and producing a powerful jet capable of rapidly extinguishing class A fires. In stored pressure extinguishers, the main body is constantly under pressure from dry air or nitrogen, and the extinguisher is operated by opening the squeeze grip discharge valve. These extinguishers are available with 6-liter- or 9-liter-capacity bodies and thus provide alternatives of weight and accessibility.

Foam is an agent most suitable for dealing with flammable liquid fires. Foam is produced when a solution of foam liquid and water is expelled under pressure through a foam-making branch pipe at which point air is entrained, converting the solution into foam. Foam extinguishers can be pressurized either by a CO_2 gas cartridge or by stored pressure. The standard capacities are 6 and 9 liters.

Spray foam. Unlike conventional foams, aqueous film forming foam (AFFF) does not need to be fully aspirated in order to extinguish fires. Spray foam extinguishers expel an AFFF solution in an atomized form that is suitable for use on class A and class B fires. AFFF is a fast and powerful means of tackling a fire and seals the surfaces of the material, preventing reignition. The capacity can be 6 or 9 liters, and operation can be by CO_2 cartridge or stored pressure.

Carbon dioxide

Designed specifically to deal with class B, class C, and electrical fire risks, these extinguishers deliver a powerful concentration of carbon dioxide gas under great pressure. This not only smothers the fire very rapidly but is also nontoxic and is harmless to most delicate mechanisms and materials.

Dry powder

This type of extinguisher is highly effective against flammable gases, as well as open or running fires involving flammable liquids such as oils, spirits, alcohols, solvents and waxes, and electrical risks. The powder is contained in the metal body of the extinguisher from which it is supplied either by a sealed gas cartridge or by dry air or nitrogen stored under pressure in the body of the extinguisher in contact with the powder.

Dry powder extinguishers are usually made in sizes containing 1 to 9 kg of either standard powder or (preferably and more generally) all-purpose powder, which is suitable for mixed risk areas.

Choosing and siting portable extinguishers. Because there is such a variety of fire risks in workshops, it is important to analyze these risks separately and (with the help of experts such as fire officers) to choose the correct firefighting medium to deal with each possible fire situation. It should be noted that portable fire extinguishers are classified as first-aid firefighting and are designed for ease of operation in an emergency. It is important to realize that because they are portable, they have only a limited discharge. Therefore, their siting, together with an appreciation of their individual characteristics, is fundamental to their success in fighting fires.

Safety signs in the workshop. It is a legal requirement that all safety signs comply with BS EN ISO 7010:2012+A7:2017.

Prohibition signs have a red circular outline and crossbar running from top left to bottom right on a white background. The symbol displayed on the sign must be black and placed centrally on the background, without obliterating the crossbar. The color red is associated with "stop" or "do not."

Warning signs have a yellow triangle with a black outline. The symbol or text used on the sign must be black and placed centrally on the background. This combination of black and yellow identifies caution.

Mandatory signs have a blue circular background. The symbol or text used must be white and placed centrally on the background. Mandatory signs indicate that a specific course of action is to be taken.

Safe condition signs provide information for a particular facility and have a green square or rectangular background to accommodate the symbol or text, which must be in white. The safety color green indicates "access" or "permission."

General safety precautions in the workshop

The Health and Safety Act is designed to ensure that:

- Employers provide a safe working environment with safety equipment and appropriate training
- Employees work in a safe manner using the equipment provided and follow the guidance and training that is provided
- Customers and others entering any business premises are safe and protected

The following section looks at some of the details related to health and safety. In all cases you should ask yourself the following questions:

- Are there any regulations relating to this, what are they, and what do I need to do to follow them?
- What is the risk involved, and how can I remove or reduce the risk?
- Is any documentation needed?

REMEMBER

It is all about keeping yourself, your colleagues, and your customers safe, as you would want them to keep you safe too.

Particular hazards may be encountered in the manufacturing or repair of motorcycles, and safety precautions associated with them are as follows:

1. Do wash before eating, drinking, or using toilet facilities to avoid transferring the residues of sealers; pigments; solvents; and steel, lead, and other metal filings from the hands to the inner parts and other sensitive areas of the body.
2. Do not use kerosene, thinners, or solvents to wash the skin. They remove the skin's natural protective oils and can cause dryness and irritation or have serious toxic effects.
3. Do not overuse waterless hand cleaners, soaps, or detergents, as they can remove the skin's protective barrier oils.
4. Always use barrier cream to protect the hands, especially against oils and greases.
5. Do follow work practices that minimize the contact of exposed skin and the length of time liquids or substances stay on the skin.
6. Do thoroughly wash contaminants such as dirty oil from the skin as soon as possible with soap and water. A waterless hand cleaner can be used when soap and water are not available. Always apply skin cream after using waterless hand cleaner.
7. Do not put contaminated or oily rags in pockets or tuck them under a belt, as this can cause continuous skin contact.
8. Do not dispose of dangerous fluids by pouring them on the ground or down drains or sewers.
9. Do not continue to wear overalls that have become badly soiled or that have acid, oil, grease, fuel, or toxic solvents spilled on them. The effect of prolonged contact from heavily soiled overalls with the skin can be cumulative and life threatening. If the soilants are or become flammable from the effect of body temperature, a spark from welding or grinding could envelop the wearer in flames with disastrous consequences.
10. Do not clean dusty overalls with an air line: it is more likely to blow the dust into the skin, with possibly serious or even fatal results.
11. Do wash contaminated or oily clothing before wearing it again.
12. Do discard contaminated shoes.
13. Wear only shoes that afford adequate protection to the feet from the effect of dropping tools and sharp and/or heavy objects on them and also from red-hot and burning materials. Sharp or hot objects could easily penetrate unsuitable footwear such as canvas plimsolls or trainers. The soles of the shoes should also be maintained in good condition to guard against upward penetration by sharp or hot pieces of metal.
14. Ensure gloves are free from holes and are clean on the inside. Always wear them when handling materials of a hazardous or toxic nature.

15. Keep goggles clean and in good condition. The front of the glasses or eyepieces can become obscured by welding spatter adhering to them. Renew the glass or goggles as necessary. Never use goggles with cracked glasses.
16. Always wear goggles when using a bench grindstone or portable grinders, disc sanders, power saws, and chisels.
17. When welding, always wear adequate eye protection for the process being used. MIG/MAG welding is particularly high in ultraviolet radiation, which can seriously affect the eyes.
18. Glasses, when worn, should have "safety" or "splinter-proof" 'glass or plastic lenses.
19. Always keep a suitable mask for use when dry flatting or working in dusty environments and when spraying adhesive, sealers, solvent-carried waxes, and paints.
20. In particularly hostile environments, such as when using volatile solvents or isocyanate materials, respirators or fresh air–fed masks must be worn.
21. Electric shock can result from the use of faulty and poorly maintained electrical equipment or misuse of equipment. All electrical equipment must be frequently checked and maintained in good condition. Flexes, cables, and plugs must not be frayed, cracked, cut, or damaged in any way. Equipment must be protected by the correctly rated fuse.
22. Use low-voltage equipment wherever possible (110 volts).
23. In case of electric shock:
    a. Avoid physical contact with the victim.
    b. Switch off the electricity.
    c. If this is not possible, drag or push the victim away from the source of the electricity using nonconductive material.
    d. Commence resuscitation if trained to do so.
    e. Summon medical assistance as soon as possible.

Electrical hazards

Voltages. The normal mains electricity voltage via a three-pin socket outlet is 240 volts; heavy-duty equipment such as machine tools use 415 volts in the form of a three-phase supply. Both 240-volt and 415-volt supplies are likely to kill anybody who touches them. Supplies of all voltages must be used through a professionally installed system and be tested regularly. If 240 volts is used for power tools, a safety circuit breaker should be used. A safer supply for power tools is 110 volts; this may be wired into the workshop as a separate circuit or provided through a safety transformer. Inspection hand-lamps are safest with a 12-volt supply.

Checklist

Before using electrical equipment, you are advised to check the following:

1. Cable condition – check for fraying, cuts, or bare wires.
2. Fuse rating – the fuse rating should be correct for the purpose as recommended by the equipment manufacturer.
3. Earth connection – all power tools must have sound earth connections.
4. Plugs and sockets – do not overload plugs and sockets; ensure that only one plug is used in one socket.
5. Water – do not use any electrical equipment in any wet conditions.
6. PAT testing – it is a requirement of the Electricity at Work Regulations that all portable electrical appliances are tested regularly; they should be marked with approved stickers and the inspection recorded in a log.

COSHH

The **Control of Substances Hazardous to Health** regulations require that assessments be made of all substances used in the workshop. This assessment must state the hazards of using the materials and how to deal with accidents arising from misuse. Your wholesale supplier will provide you with this information as set out by the manufacturer in the form of either single sheets on individual substances or a small booklet covering all the products in a range. Data sheets are available for many common products on the various manufactures' and suppliers' websites. An example can be found in the appendix.

RIDDOR

The Reporting of Injuries, Diseases and Dangerous Occurrences Regulations (RIDDOR) 2013 requires that certain information be reported to the health and safety executive (HSE). This includes the following:

- Work-related accidents that cause death
- Work-related accidents that cause certain serious injuries (reportable injuries)
- Diagnosed cases of certain industrial diseases
- Certain "dangerous occurrences" (incidents with the potential to cause harm)

Maintain the health, safety, and security of the work environment

It is the duty of every employee and employer to comply with the statutory regulations relating to health and safety and the associated guidelines that are issued by the various government offices. That means you must work in a safe and sensible manner. You are expected to follow the health and safety recommendations of your employer; employers are expected to provide a safe working environment and advise on suitable safe working methods.

Health and safety law states that organizations must:

- Provide a written health and safety policy (if they employ five or more people)
- Assess risks to employees, customers, partners, and any other people who could be affected by their activities
- Arrange for the effective planning, organization, control, monitoring, and review of preventive and protective measures
- Ensure they have access to competent health and safety advice
- Consult employees about their risks at work and current preventive and protective measures

---

Tech Note

Everybody in an organization has a duty of care related to health and safety. The HSE may bring about prosecutions or lesser prohibitions subject to timed actions – for instance, being given a short period of time to rectify a machine fault. However, the final consequences can be devastating for a firm and its employees; possible outcomes are:

- Unlimited fine
- Imprisonment
- Closing down of the business
- Disqualification from working in that job or type of business

---

Suppliers' and manufacturers' instructions relating to safety and safe use of all equipment are followed

Many pieces of equipment are marked "only to be used by authorized personnel." This is mainly because incorrect use can cause damage to the equipment, the work piece, or the operator. Do not operate equipment that you have not been properly trained to use and have not been given specific permission to use.

The suppliers of workshop equipment issue operating instructions, and as part of your training you must read these instruction booklets so that you will understand the job better. You will also find that certain safety instructions are marked on the equipment. Safe working load (SWL) in either tons or kilograms is marked on lifting equipment. You must ensure that you do not exceed these maximum load figures. Some items of equipment have two-handed controls or dead-man grips – do not attempt to operate these items incorrectly.

Approved/safe methods and techniques are used when lifting and handling

Do not attempt to manually carry a load that you cannot easily lift and that you cannot see above and around. The advised maximum weight of load that you should lift is 20 kilograms, but as a trainee this may still be too heavy for you. Do not lift weights that you are not comfortable with.

When you are lifting items from the floor, always keep your back straight and bend your knees. Bending your back while lifting can cause back injury. If you keep your feet slightly apart, this will improve your balance. It is always a good idea to wear safety gloves when manually lifting.

Required personal protective clothing and equipment are worn for designated activities and in designated areas

Table 4.1 lists typical items of personal protective equipment (PPE) and states when they must be worn.

Table 4.1 PPE Usage

PPE	Usage
Cotton overalls (boiler suit)	All the time
Safety footwear	All the time
Disposable gloves	Dealing with dirty or oily items
Rubber gloves	Operating the parts-cleaning bath
Reinforced safety gloves	Handling heavy, sharp-edged items
Dust mask	Rubbing down
Breathing apparatus and paper coveralls	Spray painting
Goggles	Using grinder or drill
Waterproof overalls and boots	Steam or pressure washing

You will often see safety notices requiring you to wear certain PPE in some areas at all times. This is because other people are working in the area and you may be at risk.

Injuries involving individuals are reported immediately to competent first-aiders and/or appropriate authorized persons, and appropriate interim support is organized to minimize further injury.

Should there be an accident, the first thing to do is call for help. Either contact your supervisor or a known first-aid person. Should any of these not be available, and it is felt appropriate, call for your local doctor or an ambulance.

You are not expected to be a first-aid expert, nor are you advised to attempt to give first aid unless you are properly qualified. However, as a professional in the motorcycle industry, you should be able to preserve the scene, that is, prevent further injury and make the injured person comfortable. The following points are suggested as ones worth remembering:

Switch off any power source.

1. Do not move the person if injury to the back or neck is suspected.
2. In the case of electric shock, turn off the electricity supply.
3. In the case of a gas leak, turn off the gas supply.
4. Do not give the person any drink or food, especially alcohol, in case surgery is needed.
5. Keep the person warm with a blanket or coat.
6. If a wound is bleeding heavily, apply pressure to the wound with a clean bandage to reduce the loss of blood.
7. If a limb has been trapped, use a safe jack to free the limb.

Visitors are alerted to potential hazards

The best policy is not to let anyone other than staff into the workshop; if entry is necessary by others, you must highlight any hazards and ensure that the company policy is complied with – usually this is done by requiring a sign-in and issuing a visitor lanyard with safety instructions.

It is always a good idea to accompany customers when they are in the workshop. This way, you can advise them in the event that they may do something potentially dangerous or if there is a hazard of which they may not be readily aware.

Injuries resulting from accidents or emergencies are reported immediately to a competent first-aider or appropriate authority

If a person is injured, the first action must be to ensure that first aid is given by a competent first-aider or other suitable person. Most companies have a designated first-aider who is trained to deal with accidents and emergencies. If your company has no such a person on the staff, then you will have a designated person who you must contact in the event of a colleague being injured. That person may be your supervisor or another senior member of the staff. If no manager or other senior person is available, you should either dial 911 for an ambulance or call your company doctor, then inform the garage manager.

Incidents and accidents are reported in an accident book

By law, all companies are required to maintain records of accidents that take place at work. These records are usually kept in an accident book. Accident books may be inspected by HSE inspectors and must be kept for a period of at least 3 years from the date of the last entry.

The information that is required to be recorded in the accident book is:

- Name and address of injured person
- Date, time, and place of accident/dangerous occurrence
- Name of person making the report and date of entry
- Brief account of accident and details of any equipment/substances that were involved

It is always a good idea to keep a notepad or smart phone to help remind you which way round things go when working on unfamiliar motorcycles or with new machinery. This would also be useful for making any other notes, such as those about an accident. Taking photographs with your phone and using the note facility is a great way of doing this.

Where there is a conflict over limitation of damage, priority is always given to the person's safety

You can always buy a new motorcycle, but you cannot buy a new arm for a mechanic. In the event of an accident, people come first. For instance, if a building is on fire, do not re-enter to retrieve your belongings; wait until the fire is out and there is no risk before going back into the building.

Reports/records are available to authorized persons and are complete and accurate

The Social Security (Claims and Payments) Regulations 1979/1987 require employers to maintain an accident book, as does Regulation 7 of the HSWA. This book requires brief details of any accident or dangerous occurrence to be recorded. An approved book BI 510 is available from the HSE direct or through most good bookstores. For more detailed information, HSE

Form 2508 should be completed. HSE staff have a statutory right to see a completed accident book or Form 2508, and they may also ask for further information. If you are personally involved in an accident, you are advised to keep a copy of the book entry and any completed forms, as well as your own notes on the event. These may be useful in the event of legal proceedings.

Machines and equipment are isolated, where appropriate, from the mains prior to cleaning and routine maintenance operations

You must always isolate an electrical machine from the mains supply before either cleaning it or carrying out any maintenance or repairs. There are two reasons for this: first, if you touch an electrically live part, you may get an electric shock; second, the machine may be accidentally started, which could cause injury or damage.

With portable electrical appliances, this simply means switching it off and taking the plug out of the socket.

With fixed machinery, for instance, a pillar drill, you will need to switch off the power supply at the isolator switch. This is usually found on the wall near the machine. Isolating this way is fine while cleaning the machine, but for carrying out maintenance or repair work, it is advisable to remove the supply fuse from the isolator box. With the fuse removed, the machine cannot be restarted if the isolator is accidentally turned on by a colleague who confuses the isolator for the one on an adjacent machine.

Safe and approved methods for cleaning machines/equipment are used

There are three main items of cleaning equipment used in the workshop: the cleaning bath (or tank), the pressure washer, and the steam cleaner.

The cleaning bath uses a chemical solvent; this is usually used for cleaning dirty/oily components. The components are submerged in the solvent, and dirt is loosened with a stiff-bristled brush.

The pressure washer is used for cleaning the mud off motorcycles; water at very high pressure will clean off mud. For hard-to-remove dirt, detergent can be added to the pressure washer. The steam cleaner, often referred to as a steam jenny (jenny = generator), produces hot pressurized water with the option of detergent. This is used for removing very stubborn grease and dirt.

If you are cleaning electrical equipment, it is important not to get water inside.

When cleaning portable electrical appliances, be careful not to get water on the plug; this could cause a short-circuit.

The mechanical parts of fixed machines may be cleaned with solvents and then dried with an absorbent paper towel.

Appropriate cleaning and sanitizing agents are used according to the manufacturer's instructions

Before using any solvent, detergent, or sanitizing agent such as bleach, you must read both the label on the container and the COSHH sheet that the manufacturer or your company has prepared.

Solvent should only be used in the cleaning bath for which it is designed.

The pressure washer or steam jenny should only be used with the recommended detergent.

Electrical items can be cleaned with one of the many aerosol sprays that are available for this purpose, but the volatile fumes that are given off must not be breathed in.

You should remember that all cleaning agents should be kept away from your mouth and eyes, and contact with your skin may cause irritation or a more serious skin disease. Always wash your hands and any other exposed areas of skin with toilet soap after carrying out a cleaning task.

Used agents are safely disposed of according to local and statutory regulations

The Environmental Protection Act (EPA) and local by-laws in most areas require that used cleaning solvents must be disposed of safely. This means that they must be put into drums and either collected by a refuse disposal firm or taken to a local authority amenity site, where they are put into a large tank for bulk incineration. Several local authorities, for instance, Surrey and Hampshire, are now looking at ways of using the energy produced by burning waste material to produce electricity. Emptying used solvents into the drain can lead to a heavy fine or even imprisonment.

Detergents are by their nature biodegradable; that is, they break down, do not build up sludge, and will not explode, unlike solvents. However, if you use large quantities of detergents, wash bays that are fitted with the correct type of drainage system should be used.

Machinery, equipment, and work areas are cleaned according to locally agreed schedules

In your company's health and safety policy document there will be reference to the cleaning of the floors and equipment in the workshop and general amenities such as toilets and rest areas. Also there will be maintenance and repair records for the workshop equipment, which will include a regular schedule of cleaning and inspection.

Most companies work on the basis of sweeping down fixed machinery and floors at the end of each day, unless the generation of dirt requires more frequent attention.

On a weekly basis there will be a more thorough cleaning program, which may include window cleaning and wet cleaning certain areas.

Workshop equipment is usually cleaned and inspected on a monthly basis unless there is reason, such as a fault, for a more regular treatment.

Appropriate safety clothing and equipment are used when working with hazardous cleansing agents and equipment

To protect yourself from the cleaning agents that you are using, you must, where appropriate, wear PPE. Most cleaning agents are poisonous and cause irritation or more serious problems if allowed to come into contact with your eyes or skin.

Whenever you are working in a workshop, it is expected that you wear cotton overalls and safety footwear. In addition the HSWA requires that employers provide and employees wear the appropriate PPE for hazardous jobs such as using cleaning equipment. The general requirements are as follows:

1. Cleaning bath – rubber protective gloves that extend over the user's wrists, goggles, and plastic apron. Avoid getting solvent on your overalls, as this can lead to skin irritation. Be especially careful not to put solvent-soaked or oily rags in your overall pockets.
2. Pressure washer – rubber protective gloves and goggles, waterproof (plastic) over-trousers and jacket, and finally rubber boots (wellingtons). The idea is to be able to take the waterproof gear off and be dry underneath.
3. Steam cleaning plant – the hazard here is that as well as being wet the water is scolding hot. So the waterproof clothes must be of such a manufacture that they will protect the wearer from the high-temperature, high-pressure steam. This means thick and strong over-trousers, coat, boots, gloves, and a hat. A full-face mask is used to give complete protection.

MISCELLANEOUS TOPICS

Lone working – Lone working is quite common in the retail and repair side of the motorcycle industry. There should be a statement in the company's health and safety policy to address this topic. A very small accident could lead to death or serious injury if the lone worker is unable to quickly call help. Keeping a mobile phone readily accessible is a good practice, with an emergency contact number on speed-dial.

First aid – The HSE recommends that in a company with 5 to 50 workers there should be at least one trained first-aider and there should be another for every 50 after that. There are different levels of first-aid training. St John Ambulance is a charity that offers first-aid training and other training and support, including online advice. They train more than 400,000 people every year.

Security – Security is a major issue with the high value and easy disposability of motorcycles. Over 15,000 motorcycles are stolen in London every year. The Metropolitan Police and many other police forces have staff just dealing with this issue. There are also issues relating to personal security and the security of premises with the ever-growing problems related to drug culture. Most police forces offer online advice and guidance on security and crime prevention.

Antitheft security – There are many antitheft security systems. There is no absolute way of preventing theft, but security marking and fitting security chips may help get a stolen motorcycle returned if found by the police. During raids on criminal suspects, many high-value objects are found each day. If yours is marked, it will lead to its return; this may also help in the conviction of the criminal by providing additional evidence.

Emergency procedures – All businesses should have emergency procedures written into their policy documents.

Emergency contacts – All businesses are advised to maintain lists of emergency contacts, a list situated near the telephone and a number next to the intruder alarm.

Ride-outs – Motorcycling club riding with colleagues requires a disciplined approach to riding to the normal journey; it only comes with practice. Many motorcycling clubs have Sunday morning ride-out events. The ride leader will usually offers a short briefing or posts the instructions online.

Insurance – Accidents and incidents do happen. Make sure that you are insured for your riding needs.

Chapter 5

Motorcycle types

The range of different types of motorcycles has grown in recent years, as manufactures have sought to meet the new types of riders. Indeed, it is now quite normal to have several motorcycles, whether this is for a specific collection or to cater for different riding needs or wants. The age and style of riders are constantly developing. Riders tend to be older, financially better off, more discerning, and more inclusive in terms of gender.

There are lots of clubs and groups for motorcycle types and marques. There are theme clubs like scooter owners and one-make clubs like Triumph and Harley Davidson owners clubs. There are groups of adventure motorcycle riders and specialist racing groups.

The driving license regulations have also brought about groups of types of motorcycles.

MOTORCYCLE CATEGORIES, AGES, AND LICENCE REQUIREMENTS IN THE UK

License category	Vehicles you can ride	Requirements for license	Minimum age
AM	Mopeds with speed range of 25 km/h to 45 km/h	Compulsory basic training (CBT), theory test, practical test on all powered two-wheeled moped	16
AM	Small three-wheelers (up to 50 cc and below 4 kW)	CBT, theory test, practical test	16
AM	Light quadricycles (weighing under 350 kg, top speed 45 km/h)	CBT, theory test, practical test	16
Q	Same as AM plus two or three-wheeled mopeds with top speed of 25 km/h	Granted with AM	16
A1	Light motorcycle up to 11 kW (and a power-to-weight ratio not more than 0.1 kW per kg) and 125 cc	CBT, theory test, practical test	17

(Continued)

License category	Vehicles you can ride	Requirements for license	Minimum age
A1	Motor tricycles with a power output not more than 15 kW	CBT, theory test, practical test	17
A2	Standard motorcycle up to 35 kW (and a power-to-weight ratio not more than 0.2 kW per kg), bike must not be derived from vehicle more than twice its power	Direct access route – theory and practical. Progressive access route – 2 years' experience on A1 motorbike and a further practical test	19
A	Unrestricted motorcycles in size/power, with or without a sidecar, and motor tricycles with power output over 15 kW	Direct access route – CBT theory and practical (you must be at least 24). Progressive access route – held an A2 license for a minimum of 2 years – practical test (21 or over)	24 (direct) or 21 (progressive access)

You do not need to take the theory or motorcycle tests to apply for a provisional license.

SAFETY EQUIPMENT

Helmet

You must wear a safety helmet when riding a motorcycle on the road. All helmets sold in the UK must comply with at least one of these:

- British Standard BS 6658:1985 and carry the BSI (British Standards Institution) Kitemark
- UNECE Regulation 22.05
- Any standard accepted by a member of the European Economic Area that offers a level of safety and protection equivalent to BS 6658:1985 and carries a mark equivalent to the BSI Kitemark

You must wear glasses or contact lenses when you ride if you need them to read a license plate at the prescribed distance.

Visors and goggles

Your visors or goggles must comply with either:

- A British Standard and displays a BSI Kitemark
- A European standard that offers a level of safety and protection at least equivalent to the British Standard and carries a mark equivalent to the BSI Kitemark (ECE 22-05)

Categories

The official categories for statistical purposes are:

- Moped
- Adventure sport
- Custom
- Naked
- Scooter
- Sport/tourer
- Supersport
- Touring
- Trial/Enduro

Mopeds are divided into two groups: 50-cc scooters and 50-cc others. "Others" refers to mopeds of different styles such as off-road and more traditional styles, tricycles, and quad-bikes. All of them are limited to 45 km/h and are usually referred to as 30 mph. These can be ridden at the age of 16 with an AM license.

Light motorcycles – Generally referred to as 125-cc class and can be ridden at the age of 17, the same as a car license. These need an A1 license.

Medium-weight motorcycles – 350-cc class; require an A2 license and the age of 19 years.

Unlimited size motorcycles – Can only be ridden over the age of 21 for experienced riders with an A2 license, or over 24 for direct entry.

These age restrictions are simply to reduce accidents. The greatest percentage of accidents, with any vehicle type, involves people under the age of 24.

Figure 5.1 Lightweight 50 cc, classed as a moped.

Motorcycle types 99

Figure 5.2 Lightweight tricycle – front view.

Figure 5.3 Lightweight tricycle – side view.

Motorcycle types 101

Figure 5.4 Triumph Rocket.

Figure 5.5 Scooter.

Motorcycle types 103

Figure 5.6 Yamaha R125.

Figure 5.7 Kawasaki Hyperbike.

Motorcycle types 105

Figure 5.8 Suzuki Naked Bike.

Figure 5.9 KTM Adventure Bike.

Figure 5.10 Red and blue Cruisers.

Motorcycle types 107

Figure 5.11 Kawasaki Ninja ZX10R.

Figure 5.12 Custom Triumph.

108 Motorcycle Engineering

Figure 5.13 Custom Triumph.

Figure 5.14 Krugger – concept bike.

Motorcycle types 109

Figure 5.15 Vyrus – concept bike.

Chapter 6

Materials for motorcycles

Motorcycles tend to be made in a traditional way, using traditional materials. This section discusses some of the materials that are used and their properties.

Engineers tend to classify materials into two major groups, each with two subgroups. The major groups are metallic materials and nonmetallic materials. We'll look at each in turn.

Motorcycle Engineering Materials

Metallic Materials	
Ferrous – contains iron	Nonferrous – does not contain iron
Iron in various forms	Aluminum
Low-carbon steel	Brass – copper and zinc
Medium-carbon steel	Bronze – copper and tin
High-carbon steel	Chromium
Alloy steel	Copper
	Titanium
Nonmetallic Materials	
Natural – occur in nature	Synthetic – manmade materials
Leather	Carbon fiber
Wood	GRP – glass fiber
Wool	Lycra/Spandex/elastane
Bamboo	Thermo-plastics
Cotton cloth	Thermo-setting plastics

METALLIC MATERIALS

The metallic group is divided into two subgroups: **ferrous metals** and **nonferrous metals**. Ferrous simply means iron – all ferrous metals contain iron. Nonferrous metals do not contain iron.

Iron is dug from the ground and heated in a furnace – there are several different types of furnaces – and mixed with carbon to form steel. Steel has been a popular choice for motorcycle construction since the start. Steel was extensively used long before motorcycles were invented. When we talk about steel, it is important to realize that there are several major categories of steel: low carbon, medium carbon, high carbon, and many types of alloy steel. When we talk about alloy steel, we simply mean that it is steel mixed with another element.

> **Tech Note**
>
> **Alloy** simply means a mixture of a metal and another element. There are alloys of steel and alloys of aluminum.

MANUFACTURE OF STEEL

Iron ore, which is dug up from the ground, is fed into a blast furnace together with limestone and coke. The coke is used as a source of heat and the limestone as a flux; that is, an agent that cleans and helps the flow of the metal. It separates the metal from the impurities in the mixture. The molten metal is now poured out of the furnace into molds to form what are called pigs – chunks of iron that resemble the shape of a pig's body. Because of the burning process, the pig iron contains between 3% and 4% carbon.

The pig iron is now changed into steel by reheating in a furnace and blasting with air to reduce the carbon content to between 0.08% and 0.20%. The term blast furnace is used, though there are other types of processes.

Casting

The steel is cast into ingots or into a continuous rolling slab, depending on the process and purpose of the material.

Pickling

The next stage is to remove the black scale from the surface of the metal. This is called pickling – the steel is run through a bath or shower of either hydrochloric acid or sulfuric acid. This ensures that the surface of the steel is clean.

Cold rolling and hot rolling

To make sheet steel, which may be converted into tubing by bending and joining, the ingots or continuous slab may be rolled. This may be done either when steel is hot or when it is cold. Rolling changes the structure of the steel and needs to be followed by annealing and tempering stages. Tubing made from sheet steel is used in the construction of some motorcycle frames.

ANNEALING AND TEMPERING

Annealing is a method of treating the steel. It is used to take away the internal stress of the steel. It needs to be stress free, or softened, to be able to be worked into the desired shape. Tempering is a process of making it the correct hardness. The annealing process usually means heating and cooling the steel in a controlled, oxygen-free atmosphere. The tempering involves reheating to a set temperature and cooling at a specific rate. As you can see, this process uses a large amount of energy – the material has been heated and cooled four times: original casting, blasting, annealing, and tempering.

Hot drawing and cold drawing

Sheet steel can be bent into tubes and seam-welded. The problem is the seam is a weakness and may tear apart under stress. Therefore, tubing made in this way tends to be made thicker to give the required factor of safety.

Tech Note

Factor of safety is the number of times that the maximum load a component can carry is divided by the expected load – this is expressed as a ratio or percentage. If a frame tube can carry a load of 650 kg before breaking and the load is 65 kg (weight of a typical rider), then the calculation is 650/65 = 10. The factor of safety is 10. This calculation is now being used in the design of racing motorcycles to get the maximum lightness. Keep in mind that dynamic stress is greater than static stress for the same load, generally called 2σ (2 sigma).

Reynolds in the UK and Columbus in Italy developed ways of drawing tubing without seams, that is, seamless tubing. Being seamless, the tubing is equally strong across its entirety.

The hot ingot is held in a die with a plug, and the tube is formed by being pulled over the plug. This process may be carried out several times to get the required tube wall thickness.

CLASSIFICATIONS OF STEEL

Steel is a highly developed product and is classified in a number of ways. General classifications are:

- Cold forming steels
- Carbon steels
- Alloy steels

- Free cutting steels
- Spring steels
- Rust-resisting and stainless steels

Cold forming steels – This is in effect sheet steel, used on pressed steel components such as petrol tanks and the headlamp nacelle. Also, on utility machines components such as brake levers and pedals are made from pressed steel, as they can be made in very large numbers very quickly. Using automated presses, a brake lever or pedal can be made in less than 5 seconds – that is up to 18,000 per day. The process for the manufacture of a component such as a headlamp nacelle requires two stages. First the material is cut to the size that is required. It is stamped out like you might cut a biscuit out of pastry. Then it is moved to a machine that will bend it – also called forming - into shape against a die. The cutters for the shape and the dies for the forming can be changed in these machines in minutes so that a factory can make nacelles in the morning and pedals in the afternoon – or any other similar product. Such companies tend to supply pressed products to a range of different industries, including the automotive industry and the construction industry.

> **SAFETY NOTE**
>
> Some motorcycle components may look and feel like they are manufactured from sheet steel, but they may not behave in a way that you would expect. For example, manufacturers in the Far East often use different materials and processes from those in Europe and the United States, so welding and other repair or modification procedures may not be possible.

Carbon steels – These are used for a large number of small motorcycle components, such as chains, gear blocks, axles and bearings, and traditional steel frames. There is a large variation in carbon steels – this leads to a set of general classifications:

- Low-carbon steel – also called mild steel – 0.10% to 0.25% carbon
- Medium-carbon steel – 0.20% to 0.50% carbon
- High-carbon steel – 0.50% to 2.00% carbon
- Tool steel
- Micro-alloyed steel

Low-carbon steel is soft, ductile, and malleable and therefore can be easily formed into shape. It cannot be hardened and tempered by heating and quenching, but it can be case-hardened and it will work-harden. Case-hardening is a process of coating the surface of the steel component with a high-carbon content chemical and heating it to a set temperature. When the component cools, the surface is hard like high-carbon steel and the underside remains soft and malleable. This process is used on hub bearing surfaces. If you look at a hub cone closely, you will be able to see the different colors of the metal. The advantages of this are that the axle and cones can be made of low-carbon steel, which is both easier to machine and cheaper to buy, and then given a wear-resistant surface for the bearing.

Low-carbon steel is usually sufficiently strong for many components on motorcycles; it is also cheap and plentiful. Most steel suppliers can offer this readily off-the-shelf.

Medium-carbon steel is much tougher and not as easy to bend or machine; it can be hardened and tempered.

Tech Note

You will find components made from carbon steel at different price points. Be aware of marketing scams – for example, the most expensive chain may not be made from the most suitable material; it may just be because it is a particular color.

ALLOYING METALS USED WITH STEEL IN MOTORCYCLE FRAMES

Chrome – A lustrous, brittle hard metal used to add corrosion resistance. It is the main additive in stainless steel. The abbreviation is **Cr**.

Manganese – Used in stainless steel to resist corrosion. Increases hardenability and tensile strength. The abbreviation is **Mn**.

Molybdenum – Used to enhance strength, improve the hardenability and weldability properties, and add toughness. It also improves corrosion resistance and high-temperature deformation. The abbreviation is **Mo**.

Vanadium – Gives added resistance to corrosion and to acids and alkalis. The abbreviation is **V**.

RESEARCH

If you are interested in researching the materials and the manufacturing of motorcycles, you'll find the following two organizations helpful:

Advanced Manufacturing Research Centre at the University of Sheffield

Advanced Materials Research Group at the University of Nottingham.

It is the author's opinion that as motorcycling is now in a new phase of development and open to new and innovative ideas, these may be either large step changes or small discreet developments.

Extrusion – Shaping components by forcing the metal through a shaped hole. The best way to explain this is to think about piping icing on to a cake. When you squeeze the icing bag, it comes out through a shaped end with a profile. Aluminum alloy motorcycle rims are extruded then curved and joined.

Hydroformed – Malleable metals such as aluminum can be formed into fairly complex shapes by hydroforming. The basic tube is put into a die assembly and then a liquid, either water based or oil based, is fed under pressure into the tube, forcing it outwards into the shape of the die.

Mar-aging steel, also written **maraging steel** without the hyphen. This word is a combination of martensitic and aging. It is a process of adding toughness and strength to low-carbon, ultra-high-strength steels by heating to a high temperature for several hours before cooling. Ultra-high-strength steels get their strength from intermetallic compounds, not added carbon. The compounds may include cobalt, molybdenum, titanium, and niobium.

Stainless precipitation hardening steel – These are low-carbon steels that have fairly high percentages of manganese, chromium, nickel, copper, and titanium. The nonferrous metals precipitate to make the steel hard. Precipitation means falling, a word that weather forecasters use for raining. In this case, it is the even distribution of these compounds of nonferrous metals in the steel, like raindrops, that makes the steel hard.

Cold worked – Means steel is rolled out when it is cold. This changes the grain structure and so makes the metal harder and stronger but reduces ductility.

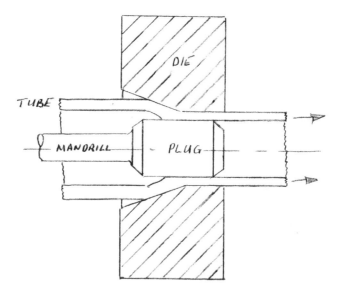

Figure 6.1 Plug and die drawing steel tube. The plug is attached to the mandrill, and the lengths of tube are pulled over the plug inside the die to form a set thickness tube.

Seamless – Tubing made from billet, not rolled and seam welded.

Air-hardening steel – This is fairly high carbon, 0.5% to 2%, with the addition of molybdenum, chromium, and manganese. It is hardened by heating to between about 800° and 900 °C then cooled in air. The heating may be carried out in a vacuum furnace.

Butting – Making the frame tubes thicker in places. Double butting is the most common – the tubes are butted at each end, where they join the other tubes.

WORK HARDENING AND FATIGUE FAILURE

We talk about hardening as a good thing, but work hardening is different; it is a bad thing. Work hardening and fatigue failure lead to component breakage. Aluminum and copper both go hard due to time and vibrations. That is, they harden without being noticeably stress loaded. Steel does not do this.

If a steel frame is not overloaded, it will retain its strength for its lifetime. An example outside the motorcycling world is the use of steel joists in buildings – these will remain straight and true if the building is not overloaded.

An aluminum frame will work-harden with time and normal road vibrations; therefore, it has a finite life span, leading to eventual fatigue failure.

Welding and brazing change the structure of the metals at the area of the joint, creating a point more susceptible to failure. The fracture usually occurs about 3 mm from the actual welded area, not of the actual weld itself.

In the author's experience, steel frames usually fail at the headstock joint and where the rear arms are attached to the wheel location plates. Aluminum frames are liable to fail at any point, depending on construction methods and usage.

PROPERTIES OF MATERIALS

Stress – This is usually measured in mega-pascals (MPa). The load in mega-newtons (mN) over the cross-sectional area in meters (m). There are several types of stress, and metals are usually judged by their ultimate tensile stress (UTS). That is the level of stress at which they will break. Making a component of thicker metal will increase the load that it can carry for any given UTS. You may use the same type metal for two frames, but if it is going to be subjected to more loads, as in motocross, then you may want to use thicker tubes; hence, the use of butting tubes at the ends.

Elongation – The amount by which something elongates, or grows longer, compared to its original length. Also, the terms **deformation** and **extension** are used where it is not a simple change in length.

Strain – The ratio of elongation divided by the original length, usually expressed as a percentage.

Young's modulus – The ratio of stress divided by strain.

Strength – Usually refers to the UTS.

Factor of safety – The number of times that the maximum load is compared to the expected load.

Elastic limit – The stress at which a metal does not return to its original shape. Steel is elastic up to a point; you bend it and it bends back. But bend it more and it stays bent.

Stiffness – The load needed to bend a tube or other component.

Strength to weight – The UTS as a ratio of the density. UTS is in MPa, density is in gram/cm^3.

Stiffness to weight – The stiffness as a ratio to density.

Aging T numbers – A set of standards that aluminum alloy is hardened to. It is expressed as a working standard, giving the temperature that the metal is heated to, the length of time it is held at this temperature, and the cooling process.

Welding and brazing dissimilar metals – The welding and brazing of dissimilar metals are possible with modern methods and fluxes. Of course, the joints and parts will have different strengths and properties from those of a normal, single-metal joint. You should check with the material suppliers and carry out a test joint before using this in a real-life situation.

EN standards – European Standards, literally European Norm. British Standards (BS), have merged with these. In America, the equivalent is ANSI, German has DIN, and Japan JIS. There are other equivalents used around the world, with hundreds of variations.

FRAME TUBING

Tubular motorcycle frames on stock machines are made from an alloy steel. On racing machines or specialist motorcycles, one of a range of high-quality steel tubing is used from one of the two major manufacturers: Reynolds tubing in the UK or Columbus in Italy.

REYNOLDS TUBING

Reynolds tubing has been used for motorcycle frames for over 100 years. Reynolds Technology invented the system of butting in 1898. They offer a range of ready-made tube sets, named and labeled as to the metal alloy that they are made from:

1. **953 Mar-aging Stainless Steel** – It combines resilience with an extremely low weight. It has high-impact strength and fatigue resistance.

2. **921 Cold Worked Stainless Steel** – High-strength, austenitic, precision-welded stainless steel. This can be shaped by frame builders without further heat treatment.
3. **853 Seamless Air Hardening Heat Treated Steel** – Light in weight with increased strength after welding. 853 is heat-treated to give high strength and damage resistance. Strong, durable, and with excellent fatigue properties, it is ideal for heavy and stronger riders.
4. **725 Heat Treated Chrome-Molybdenum Steel** – Butted and heat-treated Cr-Mo steel. A thin walled tubing gives a weight advantage over non-heat-treated tubing. Can be TIG (Tungsten Inert Gas) welded and combined in frame sets with 853 and 631 tubes.
5. **631 Seamless Air Hardened Steel** – Similar in chemistry to 853 and 631, it is cold worked and is air hardened after welding. Tough, durable, and comfortable. It can be both welded and fillet brazed.
6. **531 Manganese-Molybdenum Cold Worked Steel** – This is the grandfather of all tube sets and has been available in this format since 1935. It is perfect for brazed construction with lugs. It has a long history of use in bicycle and motorcycle frame construction. It is also used for racing motorcycle frames, racing car subframes, aircraft spars and struts, and the Trust 2 World Land Speed Record Car chassis.
7. **525 Cold Worked Chrome-Molybdenum Steel** – This is a mandrel butted frame tube made to high accuracy. It is suitable for both high-accuracy welding and fillet brazing. It has the dual advantages of being both light and competitively priced.
8. **520 Cold Worked Chrome-Molybdenum Steel** – Similar to 525 in properties. It is manufactured in Taiwan under license. It is very competitive in price and therefore suitable for large-scale production.
9. **7005 Aluminum Alloy** – This is an industry-standard aluminum alloy with zinc and magnesium elements. It is ideal for strong lightweight frames. The use of the T6 treatment process is recommended after frame construction. 7005 aluminum is also used for many applications outside the motorcycle industry.

Tech Note

T6 heat treatment condition is a procedure for ensuring that the welded aluminum frame is uniformly strong after welding. It is heated up to about 530 °C for a period of time and then cooled in a water bath. There are a number of treatments with T numbers.

6061 Aluminum Alloy – Aluminum alloyed with silicone and magnesium to provide a relatively low-cost and very lightweight material. It is very ductile and lends itself to the hydro-forming of the nonround shapes used

Table 6.1 Reynold Tubes Strength and Density (relative weight)

Tubing number	Material	UTS MPa	Density gram/cm³
953	Stainless steel	1750–2050	7.8
921	Stainless steel	950–1080	7.9
853	Steel	1200–1400	7.78
725	Chrome-molybdenum steel	1080–1280	7.78
631	Air-hardened steel	800–900	7.78
531	Manganese-molybdenum steel	650–850	7.8
525	Chrome-molybdenum steel	700–850	7.78
520	Chrome-molybdenum steel	700–900	7.78
7005	Aluminum alloy	400	2.78
6061	Aluminum alloy	325	2.7
6-4Ti	Titanium	900–1150	4.42
3-2.5Ti	Titanium	810–960	4.48

for some frames. It is readily weldable and used for a wide range of applications in other industries. It may be heat-treated using the T6 process.

6-4 Ti Seamless ELI Grade Titanium – Manufactured from custom-made billet, it is the only mandrel butted seamless 6-4 tubing. It is 6% aluminum and 4% vanadium made to ELI-grade standards. It is very lightweight, is very durable, and has the highest fatigue resistance, so it is usable for high-quality racing frames.

> **Tech Note**
>
> ELI titanium has extra-low interstitial gas purity. This means that the titanium is reduced in oxygen and iron impurities, making it stronger and less liable to fracture. ELI titanium is used for both medical purposes, such as joining fractured bones, and aerospace engineering.

3-2.5 Ti Seamless Grade Titanium – Three percent aluminum and 2.5 percent vanadium. Can be cold-worked for custom designs and is easy to weld, so it is ideal for custom-built frames and special applications. More readily available than 6-4 ELI grade tubing.

COLUMBUS TUBING

Columbus tubing is made in Milan, Italy. The company was founded by Angelo Luigi Colombo in 1919. It now constitutes two companies: Gilco and Trafiltubi. Gilco became well known for the tubing used for the chassis of Ferraris. The range of Columbus tubing currently available is:

Zonal 7000 Aluminum Alloy – This is an alloy of aluminum, zinc, and magnesium. The UTS is 420 MPa.

Spirit Alloy Steel – A very high-strength steel containing niobium. This is used for professional bicycle road racing. The UTS is 1050 to 1250 MPa.

Life Alloy Steel – Similar to Spirit.

Zona Alloy Steel – A nonsymmetrical butted tubing for top-of-the-line custom racing frames. It is alloyed chromium and molybdenum. The UTS is 800 MPa.

Megatubes – A range of oversizes and specially shaped tubes in a range of materials.

Forks in Cold-Drawn and Cold-Shaped Alloy Steel – These use pilgrim-rolling technology. Pilgrim means that one undriven roller follows the driven roller. The alloy is chrome molybdenum.

Columbus XCr – Seamless stainless steel from billet. The UTS is 1250 to 1350 MPa. A high-end frame material.

Columbus Niobium – An alloy steel with manganese, nickel, molybdenum, and niobium. The UTS is 1050 to 1150 MPa.

Columbus 25CrMo4 – A seamless alloy steel with a high chromium content. The UTS is 800 MPa.

OTHER TUBING SUPPLIERS

It should be noted that there are other tube suppliers. Reynolds and Columbus are the most popular brands because of their continued high quality. The author uses both brands, as they offer reliability and regularity in terms of both supply and performance.

Chapter 7
Frames and fairings

Frames take their shape from bicycle frames – the first motorcycles used bicycle frames; they have just developed from them.

FRAMES

The two main shapes are:
   Diamond frame – The engine and gearbox sit in the middle with the front and rear wheels attached at each end.
   Cruciform or cross frame – The wheels are at each end and the engine hangs below.

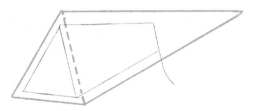

Figure 7.1 Basic diamond frame shape.

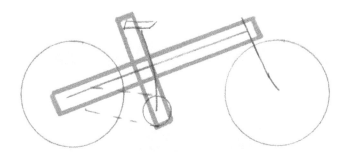

Figure 7.2 Basic cruciform frame shape.

The diamond frame is traditionally made from steel tubing brazed into lugs. Tubing is discussed in the chapter 6 on materials. Traditionally, high-quality motorcycles would use Reynolds or Columbus tubing. This type of construction has had three major variations:

- Use of fillet brazing or MIG (Metal Inert Gas) welding for the joints to reduce weight.
- Norton added an extra down-tube – called the feather-bed frame – which changed road-holding forever.
- Some manufactures removed the down-tube altogether, using the engine as a stressed member, saving weight and money.

The cruciform frame was given credence by Ducati producing the trellis frame: a cruciform frame made from short sections of thin tubing like a trellis fence. Of course, this needs a lot of welding, which takes skill and time. However, they do look beautiful and are very lightweight.

The Japanese manufacturers converted the cruciform frame into aluminum spars. This saved weight and construction time. There are three common methods of making the spars:

- An aluminum sheet is pressed to shape and welded.
- An extruded aluminum section is bent and welded.
- Cast aluminum sections are welded together.

Suzuki, Yamaha, and Kawasaki each produce about 5 million each year. The welding is machine done using jigs in manufacturing cells.

Monocoque is the name given to scooter construction. This method uses a number of pieces of sheet steel that are spot-welded together. The steel panels are pressed and stamped to shape and then spot-welded in a method similar to car body construction. This is done on a production line system. China alone produces over 10 million scooters each year.

Carbon Fiber – Surprisingly, carbon fiber is only just starting to be used for motorcycle frames. F1 cars and racing bicycles have been using it for over 30 years. The current leaders are BMW Motorrad. The problem with making frames from carbon fiber is the attachment of components – fittings have to be bonded into place. The bonding can come unstuck when subjected to heat and vibrations – something motorcycles have in abundance.

NOMENCLATURE

Motorrad – German for motorcycle – sounds good.

FAIRINGS

Original fairings and those used for record-breaking/racing motorcycles are made from glass fiber are discussed in the chapter 13 about reinforced composite materials.

They were designed purely to decrease wind resistance on both bike and rider. Current mass-produced motorcycles have fairings made from ABS or a similar plastic, which can be pressed in mass production numbers. Current fairings provide the following functions:

- Make the motorcycle aerodynamic
- Shield the rider from wind
- Provide intakes for inlet air to engine
- Provide cooling air for coolant/oil cooler
- Direct hot air away from rider

Travel at speeds over 50 mph is made very comfortable with a fairing, but for commuting at slow speeds, a naked bike is easier to use and less likely to get hot.

Figure 7.3 Lightweight motorcycle with two thin tubes.

126 Motorcycle Engineering

Figure 7.4 Part of trellis frame on Ducati.

Figure 7.5 Classic Triumph twin-spare frame.

Frames and fairings 127

Figure 7.6 BMW frame based on cruciform shape.

Figure 7.7 KTM with a mixture of cruciform and trellis.

128 Motorcycle Engineering

Figure 7.8 Heavy-duty duplex diamond-based frame.

Figure 7.9 Classic Suzuki with aluminum semi-cruciform frame.

Frames and fairings 129

Figure 7.10 Zero electric motorcycle with heavy-duty trellis frame holding a battery box.

Figure 7.11 Rear view of Zero electric motorcycle. Note that the rear arm pivot is concentric with the motor shaft. The motor is the orange/brown part.

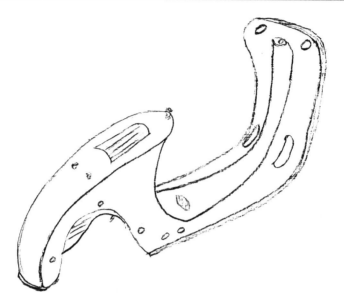

Figure 7.12 Scooter monocoque made from pressed steel sheets.

Figure 7.13 Scooter – basic monocoque with side panels.

Frames and fairings 131

Figure 7.14 BMW S100 RR. Note crash bungs and position of prop stand.

Chapter 8

Electric motorcycles

In this chapter we are going to look first at the license laws that apply to both electric bicycles and electric motorcycles. Then we discuss electric bicycles, as they are now very popular, at the same time discussing the theory of how they work. Then we look at electric motorcycles, which are starting to appear on to the roads.

> **WARNING**
>
> Do not work on electric bicycles or motorcycles unless you understand exactly what you are doing. Any voltage and any amperage of electricity can kill.

RULES AND REGULATIONS

The rules and regulations relating to electric-powered motorcycles and cycles need careful consideration to be understood. To clarify, we'll look pedal cycles first, as these are now very common.

Electrically assisted pedal cycles

> **Tech Note**
>
> This section only relates to electric bicycles, not *electric scooters, hoverboards, or personal transporters of the Segway type*. The use of these is illegal on the public roads and pavements of the UK and many other countries. The use of mobility scooters is regulated by a different set of rules.

The rules relating to the use of electric bicycles vary between both countries and types of bicycle. Looking at the UK, for example, you can ride an

134 Motorcycle Engineering

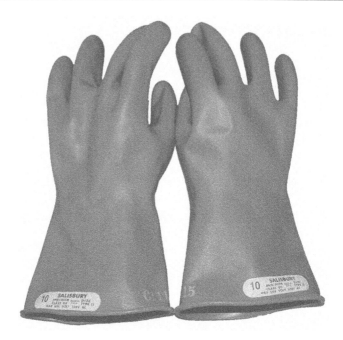

Figure 8.1 Electrically insulated gloves.

Figure 8.2 Electrically insulated tool kit – remember to keep them clean and dry.

electric bicycle in England, Scotland, and Wales if you are over 14 years old and it meets certain requirements. The electric bicycles are defined as **electrically assisted pedal cycles (EAPCs)** in the UK government rules.

By definition, an EAPC must have pedals that can be used to propel it. It must show either:

- The power output
- The manufacturer of the motor

It must also show either:

- The battery's voltage
- The maximum speed of the bicycle

Its electric motor:

- Must have a maximum power output of 250 watts
- Should not be able to propel the bike when it's traveling more than 15.5 mph

An EAPC can have more than two wheels (for example, a tricycle).

Where you can ride, if a bicycle meets the EAPC requirements, it's classed as a normal bicycle. This means you can ride it on cycle paths and anywhere else bicycles are allowed.

Any other electrically propelled bicycle that does not meet EAPC rules is classed as a motorcycle or moped and needs to be registered and taxed. You need a driving license and must wear a crash helmet to ride it. This electric bicycle must be type approved if it does not meet EAPC rules or can be propelled without pedaling, such as "twist and go."

Be aware that there is a large number of "homemade" non-EAPC electric bicycles in use. These may have been built for off-road use in the same way as Moto-X machines. The use of these is limited to restricted venues.

Still in the UK, but with different rules, in Northern Ireland, electric bicycles must be registered, taxed, and insured. In addition, the rider must have a moped license.

In France, an electric bicycle is one that does less 25 km/h (15.5 mph). EU regulations suggest that insurance is essential. In the United States, the regulations vary by state. Some states have very high cyclist casualty rates, probably attributable to high car density and the misuse of narcotics that are readily available in these areas.

Electric motorcycles

30-mph electric motorbikes

If an electric motorbike is restricted to 30 mph, the law treats it as a 30-mph 50-cc petrol scooter. They cannot be ridden on motorways.

Riders must wear a helmet, and the motorbike needs a registration document, number plate, and (once it's 3 years old) a MOT (Ministry of Transport). Electric motorbikes must be taxed, but as with all electric vehicles, road tax is free.

A 30-mph electric motorcycle can be ridden by anyone aged at least 16 with a provisional AM moped license and a Compulsory Basic Training (CBT) certificate. You must use L-plates back and front, and you cannot carry a pillion passenger. To ride without L-plates, you must take a further practical and theory test. The CBT includes both off-road and on-road riding and training. The pass certificate is valid for 2 years, or if you pass the car test in the meantime, it lasts forever.

For riders that have a car license, if your full car license was obtained before December 1, 2001, you can ride a 30-mph electric motorbike without L-plates or a CBT certificate – the pre-2001 car license automatically covers mopeds; the AM license covers petrol or electric. If the license was obtained after December 1, 2001, when the rules changed, you must have a CBT certificate first.

30-mph + electric motorbikes

There is also a power limit of 11 kw/14.6 bhp. This is in effect an A1 license; the minimum age for riding one is 17 years old. They can be ridden by car license holders with a valid CBT or a provisional A1 license and CBT.

Larger, more powerful electric motorcycles up to 35 kW require an A2 license; unrestricted power requires a full A license.

Tech Note

All motorcycles in Northern Ireland and some other countries must be insured before they can obtain a road tax or equivalent.

Only at registered off-road venues can motorcycles – electric or petrol – be ridden without tax and/or by riders without appropriate licenses.

ELECTRIC BICYCLE CONSTRUCTION

The construction of the current range of electric bicycles is based on the construction of conventional, nonelectric bicycles with the addition of three main components and a control switch/display panel. The three components are **motor, battery, and sensor.** We'll discuss each in turn.

> **Tech Note**
>
> Watt is the unit of power that has been in use since we started to use electricity. It is the product of multiplying the voltage by the amperage. In other words:
>
> Volts × Amps = Watts
>
> If you wish to compare it to your car engine, 746 watts equals 1 horsepower. Other ways of saying horsepower (HP) are cheval vapour in French (CV) and, in German, pferde stracker (PS).

Motor – The motor provides the tractive force to drive the bicycle. On an EAPC machine, the maximum allowed power output is 250 watts. That is about one-third of a horsepower. To put this into context, an average healthy male can maintain this power output for about an hour. A racing cyclist will more than double this for several hours. The power output of the motors used by the main cycle manufacturers tend to be rated slightly below the 250-watt maximum. Typically, this is 230 watts. The power output is only part of the design specification to be balanced with size, weight, and aesthetics.

Also to be borne in mind is that power can be rated in three different ways:

- Maximum power – The maximum amount that can be developed at a particular speed.
- Continuous power – This should be identified as a particular output at a particular speed for a set time.
- RMS value – The calculated average value. As a general rule:

RMS Power Value = Maximum Power × 0.7071

So, for a 250-watts maximum motor, the RMS calculation is:

RMS Power Value = 250 watts × 0.7071
176.775 watts

To make this even more complex, electric motors are often rated on their electricity consumption, not their output. The efficiency of electric motors varies considerably in their construction. Typically, the efficiency of an electric motor may be 85%; that is, 85% of the electrical power actually does useful work. So, an 85% efficient electric motor rated at 250 watts will give:

250 watts × 85 / 100 = 212.5 watts

There are three commonly used positions for the electric motor, discussed next.

Front-wheel motors

- Probably the cheapest option, as the motor fits simply in the wheel and does not affect the gears.
- It will accommodate a reasonable level of power and is ideal for city/town use.
- Good for providing steady assistance for hill climbing.
- May make the steering heavy and the machine seem a little more difficult to maneuver.
- As only part of the load of the machine is on the front wheel, there may be problems with traction in icy or wet conditions.

Rear-wheel motors

- Offers both better traction and handling, as the drive is as one would find on a normal bicycle or motorcycle.
- Has the natural feel of bicycle with a normal drive line.
- Has a stealthy appearance, just looks like a large hub gear.

Figure 8.3 Electric bicycle with bottom-bracket mounted motor.

Figure 8.4 Electric bicycle with front-wheel mounted motor.

Centre- or bottom-bracket mounted motors

- Provides a higher level of assistance, as larger motors can be used and the drive is through the gear system.
- This arrangement is better for long steep hills.
- More sensitive to your riding style, so it provides the power exactly when needed by the rider.

Battery – This stores the electrical power in chemical form. The batteries are similar to those used in industrial-standard power tools They generally work at either 24 volts, 36 volts, or 48 volts. The cell construction is usually lithium-ion (Li-ion); however, other types that are available include lead-acid (SLA), nickel-cadmium (NiCd), nickel-metal hydride (NiMh), and lithium-ion polymer (Li-pol). Li-ion batteries have the advantage of not having a memory – you do not need to fully discharge them before recharging; however, they are fairly expensive to make. Li-ion batteries are used in mobile telephones, laptop computers, and other similar devices.

There are three basic **battery ratings:**

- The nominal voltage at which they work; on electric bicycles, these are generally 24 volts, 36 volts, or 48 volts. However, be aware that these are only nominal voltages – in other words, approximate voltages. The voltage will vary with the state of charge of the battery and the load applied. When a load, for instance, the motor, is applied with a voltmeter attached across the battery terminals, you will see a drop

in the reading. This is called volt-drop and is due to the internal resistance of the battery. A faulty/old battery will have a greater volt-drop than a new one.
- The ampere-hour rating (Ah). This is not as straightforward as you would expect. A typical electric bicycle battery may be rated at 10Ah. This does not mean that it will give out 10 Amps for 1 hour, nor indeed 1 Amp for 10 hours. Batteries are tested in several different ways and averaging calculations are made.
- Watthours (Wh). This may be a fairer rating, as watts are the product of volts times amps. Typically, the batteries used on electric bicycles are about 500Wh. However, do not expect the battery to give out 500 watts for 1 hour.

> **SAFETY NOTE**
>
> There has been a great growth in battery sales. As a battery is a store of energy in chemical form, it MUST be treated with care. New European standards are coming in to use (see EN 50604).

Care of batteries – To keep your battery safe and maintain a long life, the following points should be noted:

- Store it in a cool dry place.
- Keep away from children in a secure locked cupboard.
- Keep it fully charged but not overcharged.
- Have the operating instructions to hand. Read them before using or charging the battery.
- Connect in the prescribed way to the electric bicycle, ensuring that the cable and connections are firm and secure.
- Dispose of used batteries following manufacturer's and local environmental advice.

Sensor – This is the component that switches on the motor when you start pedaling. There are two sorts of sensors:

- **Speed sensor** – This is the cheapest of the two; it switches on the motor when it senses movement and increases with rider cadence (pedaling rate). Ideal for commuter electric bicycles.
- **Torque sensor** – This increases the motor output as more pedaling effort (torque) is applied. Riding with one makes any ordinary cyclist feel like a Tour de France winner. There are several different types of torque sensors: bottom bracket, chain, crank, and derailleur hanger. They all work on the principle of measuring the torque applied by the rider – not the speed.

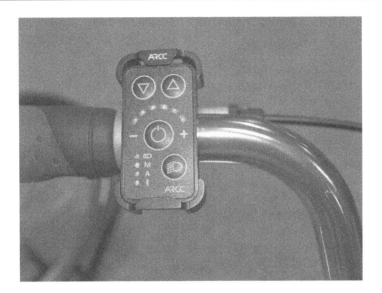

Figure 8.5 Control panel for electric bicycle.

Electric motorcycles

In terms of operation, electric motorcycles are very much akin to electric bicycles. Take a motorcycle, replace the petrol engine with an electric motor, then find somewhere to put the batteries. We now have a large variety of electric motorcycles available, but they have two problems that are causing market resistance in terms of sales: very high price tag and limited range. However, they are improving all the time.

The electric motorcycles at the Isle of Man TT are certainly at the cutting edge of this technology. At the inaugural TT Zero, one lap for electric motorcycles was won with a lap speed of 96.82 mph – this has risen to 120 mph. Petrol motorcycles are averaging 130 mph for four laps. The electric machines are typically 50% heavier than their petrol-propelled cousins. The problem is the batteries. It's good to remember that the first mobile phones were the size of an old-fashioned telephone hand-set, but the batteries filled a small suitcase. The batteries for electric motorcycles are usually lithium based and made up of cells based on 3.2-volt, 3.6-volt, or 3.7-volt output. These are connected to give out either 72 volts or 96 volts. So, we are talking seriously high voltages backed up with a house-fuse blowing 20 amps. Team Mugen – run by Mr. Honda junior – has won the TT Zero since 2014. I suspect that the aim is to beat the petrol-powered machines.

SAFETY NOTE

Ninety-six volts at 20 amp is going to make a big spark – like you get off an arc welder. This could kill you.

It would appear that the issues to be resolved by electric motorcycle makers are:

- Reduce overall weight
- Increase the distance that can be traveled on a battery charge
- Develop some form of self-charging like Lexus cars
- Improve quick charging battery life
- Develop a foolproof motor speed and torque control system

HEALTH AND SAFETY

The hazard when working on a broken-down electric bicycle or motorcycle is that you don't know if there is a short-circuit from which you could get an electric shock. Also bear in mind that there could be electrically charged components that need discharging. Three safety precautions that you must follow are:

- Always disconnect the battery when it is safe to do so.
- Always wear certified electrically insulated gloves.
- Always use certified electrically insulated tools.

Figure 8.6 Zero electric motorcycle. The large gray part under what looks like the tank – which actually houses some of the electronics – is the battery. The orange/brown component is the electric motor; this directly drives the rear wheel with a belt.

Chapter 9
Soldering, brazing, and welding

Motorcycle construction involves components being joined together. Soldering, brazing, and welding are three common methods of joining. In this chapter we discuss these processes.

COMPARISON OF FUSION AND NONFUSION JOINTING PROCESSES

The joining of metals by processes employing fusion of some kind – that is the melting of metal. There are different types of fusion. They may be classified as follows:

Total fusion

Temperature range: Approximately 1130° to 1550 °C. Processes: Oxy-acetylene welding, manual metal arc welding, inert gas metal arc welding. A welded frame, usually without lugs.

Skin fusion

Temperature range: Approximately 620° to 950 °C. Processes: Flame brazing, silver soldering, aluminum brazing, bronze welding. A lugged frame or one that is fillet brazed.

Surface fusion

Temperature range: Approximately 183° to 310 °C. Process: Soft soldering. Used on lightly loaded components and for electrical and electronic components.

In total fusion the parent metal and, if used, the filler metal are both completely melted during the jointing. Tubes can be fused together without additional filler metal being added. Oxy-acetylene welding and manual metal arc welding were the first processes to employ total fusion. In recent years they have been supplemented by methods such as inert-gas arc welding, metal inert-gas (MIG/MAG), and tungsten inert-gas (TIG) welding; carbon dioxide welding; and atomic hydrogen welding. Welding is normally carried out at high temperature ranges, with the actual temperature being the melting point of the particular metal which is being joined. The parent metal is totally melted throughout its thickness, and in some cases molten filler metal of the correct composition is added by means of rods or consumable electrodes of convenient size. A neat reinforcement weld bead is usually left protruding above the surface of the parent metal, as this gives good mechanical properties in the completed weld. Most metals and alloys can be welded effectively, but there are certain exceptions which, because of their physical properties, are best joined by alternative methods.

In skin fusion the skin or surface grain structure only of the parent metal is fused to allow the molten filler metal to form an alloy with the parent metal. Hard solders are used in this process, and as these have greater shear strength than tensile strength, the tensile strength of the joint must be increased by increasing the total surface area between the metals. The simplest method of achieving this is by using a lapped joint in which the molten metal flows between the adjoining surfaces. This accounts for the use of lugs in joining frame tubes. The strength of the joint will be dependent upon the wetted area between the parts to be joined. Skin fusion has several advantages. First, since the filler metals used in these processes have melting points lower than the parent metal to which they are being applied, a lower level of heat is needed than in total fusion, and as a consequence, distortion is reduced. Second, dissimilar metals can be joined by applying the correct amount of heat to each parent metal, when the skins of both will form an alloy with the molten hard solder. Frame lugs are usually cast from a different alloy of steel from that of the frame tubes. Third, since only the skin of the parent metal is fused, a capillary gap is formed in the lap joint, and the molten filler metal is drawn into the space between the surfaces of the metals. This allows easy assembly and wiggle room to adjust the frame angles before making the joint permanent. The filler material, also called spelter, will fill the space available when at the correct temperature to flow.

In surface fusion the depth of penetration of the molten solder into the surfaces to be joined is so shallow that it forms an intermetallic layer that bonds the surfaces together. The process employs soft solders, which are composed mainly of lead and tin. As these also have a low resistance to a tensile pulling force, the joint design must be similar to that of the skin fusion process (i.e., a lapped joint).

Soft and hard solders

In spite of the growing use of welding, the technique of soldering remains one of the most familiar in the fabrication of sheet metal articles, and the art of soldering still continues to occupy an important place in the workshop.

While soldering is comparatively simple, it requires care and skill and can only be learned by actual experience.

Soldering and brazing are methods of joining components by lapping them together and using a low-melting-point alloy so that the parent material is not melted. Soldering as a means of joining metal sheets has the advantage of simplicity in apparatus and manipulation, and with suitable modifications it can be applied to practically all commercial metals.

Soft soldering

The mechanical strength of soft-soldered sheet metal joints is usually on the order of 15 to 30 MN/m^2 and depends largely upon the nature of the solder used; the temperature at which the soldering is done; the depth of penetration of the solder, which in turn depends on capillary attraction (i.e., on the power of the heated surface to draw liquid metal through itself); the proper use of correctly designed soldering tools; the use of suitable fluxes; the speed of soldering; and, especially, workmanship.

Solders

Soft solder is an alloy of lead and tin, and is used with the aid of a soldering flux. It is made from two base metals: tin and lead. Tin has a melting point of 232 °C and lead 327 °C, but the alloy has a lower melting point than either of the two base metals, and its lowest melting point is 18 3°C; this melting point may be raised by varying the percentage of lead or tin in the alloy. A small quantity of antimony is sometimes used in soft solder with a view to increasing its tenacity and improving its appearance by brightening the color. The small percentage of antimony both improves the chemical properties of the solder and increases its tensile strength, without appreciably affecting its melting point or working properties.

> ### Tech Note
> There is a great variety of solders: aluminum, bismuth, cadmium, silver, gold, pewterer's, plumber's, tinman's; solders are usually named according to the purpose for which they are intended.

The following solders are the most popular in use today:

95% to 100% tin solder, is used for high-quality electrical work where maximum electrical conductivity is required, since the conductivity of pure tin is 20% to 40% higher than that of the most commonly used solders.

60/39.5/0.5 (tin/lead/antimony) solder, the eutectic composition, has the lowest melting point of all tin–lead solders and is quick setting. It also has the maximum bulk strength of all tin–lead solders and is used for fine electrical and tinsmith's work.

50/47.5/2.5 (tin/lead/antimony) solder, called tinman's fine, contains more lead and is therefore cheaper than the 60/40 grade. Its properties in terms of low melting range and quick setting are still adequate, and hence it is used for general applications.

45/52.5/2.5 (tin/lead/antimony) solder, known as tinman's soft, is cheaper because of the higher lead content but has poorer wetting and mechanical properties. This solder is widely used for can soldering, where maximum economy is required, and for any material that has already been tin plated so that the inferior wetting properties of the solder are not critical.

30/68.5/1.5 (tin/lead/antimony) solder, known as plumber's solder, is also extensively used by the motorcycle body repairer. Because the material has a wide liquidus–solidus range (about 80 °C), it remains in a pasty form for an appreciable time during cooling, and while in this condition it can be shaped or "wiped" to form a lead pipe joint or to the shape required for filling dents in frame tubes. Because of its high lead content, its wetting properties are very inferior, and the surfaces usually have to be tinned with a solder of higher tin content first.

Fluxes

The function of a flux is to remove oxides and tarnish from the metal to be joined so that the solder will flow, penetrate, and bond to the metal surface, forming a good strong soldered joint. The hotter the metal, the more rapidly the oxide film forms. Without the chemical action of the flux on the metal, the solder would not tin the surface, and the joint would be weak and unreliable. As well as cleaning the metal, flux ensures that no further oxidation from the atmosphere that could be harmful to the joint takes place during soldering, as this would restrict the flow of soldering.

Generally, soft-soldering fluxes are divided into two main classes: corrosive fluxes and noncorrosive fluxes.

Tech Note

Some fizzy drinks contain phosphoric acid – if you drop a dirty coin in the drink, the acid will clean it.

Brazing

Brazing is used extensively throughout the frame building trade as a quick and cheap means of joining frame tubes and other components. Although a brazed joint is not as strong as a fusion weld, it has many advantages that make it useful for the frame builder. Brazing is not classed as a fusion process, and therefore cannot be called welding, because the parent metals are not melted to form the joint, but rely on a filler material of a different metal of low melting point, which is drawn through the joint. The parent metals can be similar or dissimilar as long as the alloy rod has a lower melting point than either of them. The most commonly used alloy is of copper and zinc, which is, of course, brass. Brazing is accomplished by heating the pieces to be joined to a temperature higher than the melting point of the brazing alloy (brass). With the aid of flux, the melted alloy flows between the parts to be joined due to capillary attraction and actually diffuses into the surface of the metal, so that a strong joint is produced when the alloy cools. Brazing, or hard soldering to give it its proper name, is in fact part fusion and is classed as a skin fusion process.

Brazing is carried out at a much higher temperature than that required for the soft soldering process. A borax type of powder flux is used, which fuses to allow brazing to take place between 750° and 900 °C. There are a wide variety of alloys in use as brazing rods; the most popular compositions contain copper in the ranges of 46% to 50% and 58.5% to 61.5%, with the remaining percentage being zinc.

The brazing process comprises the following steps:

1. Thoroughly clean the metal to be joined.
2. Using a welding torch, heat the metals to a temperature below their own critical or melting temperature. In the case of steel, the metal is heated to a dull cherry red.
3. Apply borax flux either to the rod or to the work as the brazing proceeds to reduce oxidation and to float the oxides to the surface.
4. Use the oxy-acetylene torch with a neutral flame, as this will give good results under normal conditions.

SAFETY NOTE

When heated zinc-plated steel (galvanized) gives off very toxic fumes, so full respiratory equipment must be used – it is better to avoid this hazard if possible.

The main advantages of brazing are:

1. The relatively low temperature (750° to 900 °C) necessary for a successful brazing job reduces the risk of distortion.
2. The joint can be made quickly and neatly, requiring very little cleaning up.
3. Brazing makes possible the joining of two dissimilar metals; for example, brass can be joined to steel.

4. It can be used to repair parts that have to be rechromed. For instance, a chromed fork that has been deeply scratched can be readily filled with brazing and then filed up, ready for chroming.
5. Brazing is useful for joining steels that have a high carbon content, or broken castings where the correct filler rod is not available.

Silver soldering

Silver solder probably originated in the manufacture and repair of silverware and jewelry for the purpose of securing adequate strength and the desired color of the joint; but the technique of joining sheet metal products and components with silver solder was used for a long time on high-quality bicycle frames. The term "soft soldering" has been widely adopted when referring to the older process to avoid confusion with the newer hard soldering process, known generally as either silver soldering or silver brazing. The use of silver solder on metals and alloys other than silver has grown largely because of the perfection by manufacturers of these solders, which makes them easily applicable to many metals and alloys by means of the oxy-acetylene welding torch.

Solders and fluxes

Silver solders are more malleable and ductile than brazing rods, and hence joints made with silver solder have a greater resistance to bending stresses, shocks, and vibration than those made with ordinary brazing alloys. As you can see this is very appropriate for bicycle frames. Silver solders are made in strip, wire (rod), or granular form and in a number of different grades of fusibility. The melting points vary between 630° and 800 °C according to the percentages of silver, copper, zinc, and cadmium they contain.

As in all nonfusion processes the important factor is that the joint to be soldered must be perfectly clean. Hence, special care must be taken in preparing the metal surfaces to be joined with silver solder. Although fluxes will dissolve films of oxide during the soldering operation, frame tubes and lugs that are clean are much more likely to make a stronger, sounder joint than when impurities are present. The joints should fit closely, and the parts must be held together firmly while being silver soldered, because silver solders in the molten state are remarkably fluid and can penetrate into minute spaces between the metals to be joined. The use of a frame building jig is essential with this process. In order to protect the metal surface against oxidation and to increase the flowing properties of the solder, a suitable flux such as borax or boric acid is used.

Silver soldering process

In silver soldering the size of the welding tip used and the adjustment of the flame are very important to avoid overheating, as prolonged heating promotes oxide films, which weaken both the base metal and the joint material. This should be guarded against by keeping the luminous cone of the flame well back from the point being heated. When the joint has been heated just above the temperature at which the silver solder flows, the flame should be moved away and the solder applied to the joint, usually in rod form. The flame should then be played over the joint so that the solder and flux flow freely through the joint by capillary attraction. The finished silver soldered joint should be smooth, regular in shape, and require no dressing up apart from the removal of the flux by washing in water.

When making a silver solder joint between dissimilar metals, concentrate the application of heat on the metal that has the higher heat capacity. This depends on the thickness and the thermal conductivity of the metals. The aim is to heat both members of the joint evenly so that they reach the soldering temperature at the same time.

The most important points during silver soldering are:

1. Cleanness of the joint surfaces
2. Use of the correct flux
3. The avoidance of overheating.

ALUMINIUM BRAZING

There is a distinction between the brazing of aluminum and the brazing of other metals. For aluminum, the brazing alloy is one of the aluminum alloys having a melting point below that of the parent metal. For other metals, the brazing alloys are often based on copper–zinc alloys (brasses – hence the term brazing) and are necessarily dissimilar in composition to the parent metal.

Wetting and fluxing

When a surface is wetted by a liquid, a continuous film of the liquid remains on the surface after draining. This condition, essential for brazing, arises when there is mutual attraction between the liquid flux and solid metal due to a form of chemical affinity. Having accomplished its primary duty of removing the oxide film, the cleansing action of the flux restores the free affinities at the surface of the joint faces, promoting wetting by reducing the contact angle developed between the molten brazing alloy and parent metal. This action assists spreading and the feeding of brazing alloy to the capillary spaces, leading to the production of well-filled joints. An important feature of the brazing process is that the brazing alloy is drawn into the joint area by capillary attraction: the smaller the gap between the two metal faces to be joined, the deeper the capillary penetration.

The various grades of pure aluminum and certain alloys are amenable to brazing. Aluminum– magnesium alloys containing more than 2% magnesium are difficult to braze, as the oxide film is tenacious and hard to remove with ordinary brazing fluxes. Other alloys cannot be brazed because they start to melt at temperatures below that of any available brazing alloy. Aluminum–silicon alloys of nominal 5%, 7.5%, or 10% silicon content are used for brazing aluminum and the alloy of aluminum and 1.5% manganese.

The properties required for an effective flux for brazing aluminum and its alloys are as follows:

1. The flux must remove the oxide coating present on the surfaces to be joined. It is always important that the flux be suitable for the parent metal, but especially so in the joining of aluminum–magnesium alloys.
2. It must thoroughly wet the surfaces to be joined so that the filler metal may spread evenly and continuously.
3. It must flow freely at a temperature just below the melting point of the filler metal.
4. Its density when molten must be lower than that of the brazing alloy.
5. It must not attack the parent surfaces dangerously in the time between its application and removal.
6. It must be easy to remove from the brazed assembly.

Many types of proprietary fluxes are available for brazing aluminum. These are generally of the alkali halide type, which are basically mixtures of the alkali metal chlorides and fluorides. Fluxes and their residues are highly corrosive and therefore must be completely removed after brazing by washing with hot water.

Brazing method

When the cleaned parts have been assembled, brazing flux is applied evenly over the joint surface of both parts to be brazed and the filler rod (brazing alloy). The flame is then played uniformly over the joint until the flux has dried and become first powdery, then molten and transparent. (At the powdery stage care is needed to avoid dislodging the flux, and it is often preferable to apply flux with the filler rod.) When the flux is molten the brazing alloy is applied, preferably from above, so that gravity assists in the flow of metal. In good practice the brazing alloy is melted by the heat of the assembly rather than directly by the torch flame. Periodically the filler rod is lifted and the flame is used to sweep the liquid metal along the joint; but if the metal is run too quickly in this way it may begin to solidify before it properly diffuses into the mating surfaces. Trial will show whether more than one feed point for the brazing alloy is necessary, but with proper fluxing, giving an unbroken path of flux over the whole joint width, a single feed is usually sufficient.

Bronze welding

Bronze welding is carried out much as in fusion welding except that the base metal is not melted. The base metal is simply brought up to tinning temperature (dull red color), and a bead is deposited over the seam with a bronze filler rod. Although the base metal is never actually melted, the unique characteristics of the bond formed by the bronze rod are such that the results are often comparable to those secured through fusion welding. Bronze welding resembles brazing, but only up to a point. The application of brazing is generally limited to joints where a close fit or mechanical fastening serves to consolidate the assembly and the joint is merely strengthened or protected by the brazing material. In bronze welding the filler metal alone provides the joint strength, and it is applied by the manipulation of a heating flame in the same manner as in gas fusion welding. The heating flame is made to serve the dual purpose of melting off the bronze rod and simultaneously heating the surface to be joined. The operator in this manner controls the work: hence the term "bronze welding."

Almost any copper–zinc alloy or copper–tin alloy or copper–phosphorus alloy can be used as a medium for such welding, but the consideration of costs, flowing qualities, strength, and ductility of the deposit have led to the adoption of one general-purpose 60–40 copper–zinc alloy with minor constituents incorporated to prevent zinc oxide forming and to improve fluidity and strength. Silicon is the most important of these minor constituents, and its usefulness is apparent in three directions. First, in the manner with which it readily unites with oxygen to form silica, silicon provides a covering for the molten metal which prevents zinc volatilization and serves to maintain the balance of the constituents of the alloy; this permits the original high strength of the alloy to be carried through to the deposit. Second, this coating of silica combines with the flux used in bronze welding to form a very fusible slag, and this materially assists the tinning operation, which is an essential feature of any bronze welding process. Third, by its capacity for retaining gases in solution during solidification of the alloy, silicon prevents the formation of gas holes and porosity in the deposited metal, which would naturally reflect unfavorably upon its strength as a weld.

It is essential to use an efficient and correct flux. The objects of a flux are first, to remove oxide from the edges of the metal, giving a chemically clean surface on to which the bronze will flow and to protect the heated edges from the oxygen in the atmosphere; second, to float oxide and impurities introduced into the molten pool to the surface, where they can do no harm. Although general-purpose fluxes are available, it is always desirable to use the fluxes recommended by the manufacturer of the particular rod being employed.

Bronze welding procedure

1. An essential factor for bronze welding is a clean metal surface. If the bronze is to provide a strong bond, it must flow smoothly and evenly over the entire weld area. Clean the surfaces thoroughly with a stiff wire brush. Remove all scale, dirt, or grease; otherwise, the bronze will not adhere. If a surface has oil or grease on it, remove these substances by heating the area to a bright red color and thus burning them off.
2. On thick sections, especially in repairing castings, bevel the edges to form a 90-degree V-groove. This can be done by chipping, machining, filing, or grinding.
3. Adjust the torch to obtain a slightly oxidizing flame. Then heat the surfaces of the weld area.
4. Heat the bronzing rod and dip it in the flux. (This step is not necessary if the rods have been prefluxed.) In heating the rod, do not apply the inner cone of the flame directly to the rod.
5. Concentrate the flame on the starting end until the metal begins to turn red. Melt a little bronze rod on to the surface and allow it to spread along the entire seam. The flow of this thin film of bronze is known as the tinning operation. Unless the surfaces are tinned properly, the bronzing procedure to follow cannot be carried out successfully. If the base metal is too hot, the bronze will tend to bubble or run around like drops of water on a warm stove. If the bronze forms into balls that tend to roll off, just as water would if placed on a greasy surface, then the base metal is not hot enough. When the metal is at the proper temperature the bronze spreads out evenly over the metal.
6. Once the base metal is tinned sufficiently, start depositing the proper size beads over the seam. Use a slightly circular torch motion, and run the beads as in regular fusion welding with a filler rod. Keep dipping the rod in the flux as the weld progresses forward. Be sure that the base metal is never permitted to get too hot.
7. If the pieces to be welded are grooved, use several passes to fill the V. On the first pass make certain that the tinning action takes place along the entire bottom surface of the V and about halfway up on each side. The number of passes to be made will depend on the depth of the V. When depositing several layers of beads, be sure that each layer is fused into the previous one.

HEALTH AND SAFETY AND THE ENVIRONMENT

The materials used in soldering can be dangerous if not used in a safe way. In addition to the normal workshop precautions, the following should be especially noted:

- All soldering and brazing must be carried out in an area with suitable fume extraction.
- Fluxes are mainly corrosive and should be handled accordingly.
- Always wash your hands after soldering or brazing.
- Use personal protective equipment (PPE) as directed by your company safety guidelines.
- All materials must be stored securely and not be accessible to unauthorized personnel.

SAFETY NOTE

You must remember that all gases are dangerous if not handled and stored correctly – always follow the manufacturer's instructions.

The following is a summary of gas characteristics and cylinder color codes.

Oxygen

Cylinder color: Black.

Characteristics: No smell. Generally considered nontoxic at atmospheric pressure. Will not burn but supports and accelerates combustion. Materials not normally considered combustible may be ignited by sparks in oxygen-rich atmospheres.

Nitrogen

Cylinder color: Gray with black shoulder. Characteristics: No smell. Does not burn. Inert, so will cause asphyxiation in high concentrations.

Argon

Cylinder color: Blue.

Characteristics: No smell. Heavier than air. Does not burn. Inert. Will cause asphyxiation in absence of sufficient oxygen to support life. Will readily collect in the bottom of a confined area. At high concentrations, almost instant unconsciousness may occur followed by death. The prime danger is that there will be no warning signs before unconsciousness occurs.

Propane

Cylinder color: Bright red and bearing the words "Propane" and "Highly flammable."

Characteristics: Distinctive fish-like, offensive smell. Will ignite and burn instantly from a spark or piece of hot metal. It is heavier than air and will collect in ducts, drains, or confined areas. Fire and explosion hazard.

Acetylene

Cylinder color: Maroon.

Characteristics: Distinctive garlic smell. Fire and explosion hazard. Will ignite and burn instantly from a spark or piece of hot metal. It is lighter than air and less likely than propane to collect in confined areas. Requires minimum energy to ignite in air or oxygen. Never use copper or alloys containing more than 70% copper or 43% silver with acetylene.

Hydrogen

Cylinder color: Bright red.

Characteristics: No smell. Nontoxic. Much lighter than air. Will collect at the highest point in any enclosed space unless ventilated there. Fire and explosion hazard. Very low ignition energy.

Carbon dioxide

Cylinder color: Black or black with two vertical white lines for liquid withdrawal.

Characteristics: No smell but can cause the nose to sting. Harmful. Will cause asphyxiation. Much heavier than air. Will collect in confined areas.

Argoshield

Cylinder color: Blue with green central band and green shoulder.

Characteristics: No smell. Heavier than air. Does not burn. Will cause asphyxiation in absence of sufficient oxygen to support life. Will readily collect at the bottom of confined areas.

SAFETY MEASURES

General gas storage procedures

1. Any person in charge of storage of compressed gas cylinders should know the regulations covering highly flammable liquids and compressed gas cylinders, as well as the characteristics and hazards associated with individual gases.
2. It is best to store both full and empty compressed gas cylinders in the open, in a securely fenced compound, but with some weather protection.
3. Within the storage area oxygen should be stored at least 3 m from the fuel gas supply.

4. Full cylinders should be stored separately from the empties, and cylinders of different gases, whether full or empty, should be segregated from each other.
5. Other products must not be stored in a gas store, particularly oils or corrosive liquids.
6. It is best to store all cylinders upright, taking steps, particularly with round-bottomed cylinders, to see that they are secured to prevent them from falling. Acetylene and propane must *never* be stacked horizontally in storage or in use.
7. Storage arrangements should ensure adequate rotation of stock.

Acetylene cylinders

1. The gas is stored together with a solvent (acetone) in maroon-painted cylinders, at a pressure of 17.7 bar maximum at 15 °C. The cylinder valve outlet is screwed left-handed.
2. The hourly rate of withdrawal from the cylinder must not exceed 20% of its content.
3. Pressure gauges should be calibrated up to 40.0 bar.
4. As the gas is highly flammable, all joints must be checked for leaks using soapy water.
5. Acetylene cylinders must be stored and used in an upright position and protected from excessive heat and coldness.
6. Acetylene can form explosive compounds in contact with certain metals and alloys, especially those of copper and silver. Joint fittings made of copper should not be used under any circumstances.
7. The color of cylinders, valve threads, or markings must not be altered or tampered with in any way.

Oxygen cylinders

1. This gas is stored in black-painted cylinders at a pressure of 200/230 bar maximum at 15 °C.
2. Never under any circumstances allow oil or greases to come into contact with oxygen fittings because spontaneous ignition may take place.
3. Oxygen must not be used in place of compressed air.
4. Oxygen escaping from a leaking hose will form an explosive mixture with oil or grease.
5. Do not allow cylinders to come into contact with electricity.
6. Do not use cylinders as rollers or supports.
7. Cylinders must not be handled roughly, knocked, or allowed to fall to the ground.

GENERAL EQUIPMENT SAFETY

All equipment should be subjected to regular periodic examination and overhaul. Failure to do so may allow equipment to be used in a faulty state and may be dangerous.

Rubber hose. Use only hoses in good condition, fitted with the special hose connections attached by permanent ferrules. Do not expose the hose to heat, traffic, slag, sparks from welding operations, or oil or grease. Renew the hose as soon as it shows any sign of damage.

Pressure regulators. Always treat a regulator carefully. Do not expose it to knocks, jars, or sudden pressure caused by rapid opening of the cylinder valve. When shutting down, release the pressure on the control spring after the pressure in the hoses has been released. Never use a regulator on any gas except that for which it was designed, or for higher working pressures. Do not use regulators with broken gauges.

Welding torch. When lighting up and extinguishing the welding torch, the manufacturer's instructions should always be followed. To clean the nozzle, use special nozzle cleaners, never a steel wire.

Fluxes. Always use welding fluxes in a well-ventilated area.

Goggles. These should be worn at all times during welding, cutting, or merely observing.

Protection. Leather or fire-resistant clothing should be worn for all heavy welding or cutting. The feet should be protected from sparks, slag, or cut material falling on them.

GAS-SHIELDED ARC WELDING (MIG, MAG, AND TIG)

Development of gas-shielded arc welding

Originally the process evolved in America in 1940 for welding in the aircraft industry. It developed into the tungsten inert-gas shielded arc process, which in turn led to shielded inert-gas metal arc welding. The latter became established in the UK in 1952.

In the gas-shielded arc process, heat is produced by the fusion of an electric arc maintained between the end of a metal electrode, either consumable or nonconsumable, and the part to be welded, with a shield of protective gas surrounding the arc and the weld region. There are at present in use three different types of gas-shielded arc welding:

Tungsten inert gas (TIG). The arc is struck by a nonconsumable tungsten electrode and the metal to be welded, and filler metal is added by feeding a rod by hand into the molten pool.

Metal inert gas (MIG). This process employs a continuous feed electrode, which is melted in the intense arc heat and deposited as weld metal: hence the term consumable electrode. This process uses only inert gases, such as argon and helium, to create the shielding around the arc.

Metal active gas (MAG). This is the same as MIG except that the gases have an active effect upon the arc and are not simply an inert envelope. The gases used are carbon dioxide or argon/carbon dioxide mixtures.

Tech Note

Gas tungsten arc welding (GTAW) is the terminology used in America and many parts of Europe for the TIG welding process, and it is becoming increasingly accepted as the standard terminology.

Chapter 10

Running gear and braking

RUNNING GEAR

Running gear is the name given to the remains of the motorcycle that are attached to the frame when the engine and gear box have been removed. It's not specific, nor definitive.

The VIN number should enable you to identify it as to make and model. Motorcycles are notorious for not having the correct VIN number or an engine number that does not match. You will see advertisements for classic motorcycles declaring matching numbers – for a collector this is very important. There are many reasons why VIN, frame, and engine numbers may give you problems; some of them are:

- New engine or one from another machine has been fitted as part of a repair.
- About 40,000 motorcycles are stolen each year in the UK – yes almost 1,000 per week, so lots of engines, frames, and VIN/other numbers not with their owners.
- Rebuilding write-off machines using numbers on parts.
- Not all new motorcycles are given VIN numbers – one reason for this is that the manufacturers would avoid having to pay tax on them.
- Of course, you can build one up using spare parts too.

SUSPENSION AND STEERING

Most motorcycles have telescopic front forks and a single rear shock absorber with a coil spring. This is both cheap and light – the essential criteria for suspension.

Function of the suspension mechanism

The function of the suspension is to connect the road wheels to the frame; it is designed to prevent the bumps caused by the road-surface irregularities

160   Motorcycle Engineering

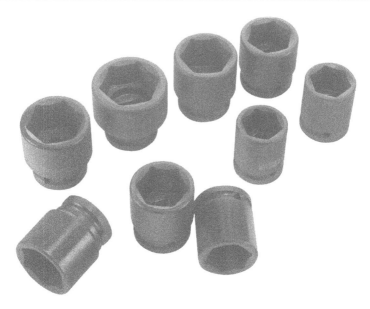

Figure 10.1 Hexagonal impact sockets – good for use on wheels and suspension, as they can withstand high loads.

from reaching the rider. This is to make the bike both pleasant to ride in and safe on the road. The suspension also protects the mechanical components from road vibrations, thus making them last much longer.

The suspension consists of a system of movable linkages and a spring and a damper (also called a shock absorber) for each wheel. The tire also forms part of the suspension; the flexing of the tire absorbs small irregularities in

Figure 10.2 Radially spoked wheel.

Figure 10.3 Cast aluminum wheel.

the road and helps keep the noise to a minimum. You might imagine what it was like before motorcycles were fitted with pneumatic rubber tires.

Castor

Castor, or to give its full name, the castor angle, is the angle that the forks lean backwards in the longitudinal plane. By giving the forks a castor angle of between 20 and 30 degrees, the imaginary center line meets the road before the center of the wheel. This distance is called the castor trail. The castor angle gives the front wheel a self-aligning, or self-centering, action. The wheel follows the pivot, thus keeping the bike on a straight course and reducing the need to manually straighten the steering wheel after negotiating a corner. This can be likened to the castors on a trolley. The castors swing round so that the trolley can be pushed in a straight line. You can see this in action on any supermarket trolley.

Coil spring

The coil spring is usually fitted around the shock absorber. It is made from round section spring steel that is wound to shape. Coil springs offer a large amount of suspension movement and lightness – ideal characteristics in most cases.

Leaf spring

These are made from a number of flat metal sections – rather like thin leaves of a book. The main leaf has an eye at each end so that it can be attached to the frame. Leaf springs can carry a lot of weight, but suspension

162 Motorcycle Engineering

Figure 10.4 Large castor angles on these two cruisers.

Figure 10.5 Lightweight coil over shock absorber.

travel is limited. In some instances, only one end is connected to the frame; the other end is free to slide. They are occasionally found on sidecars and specialized motorcycles.

Torsion bar spring

The torsion bar is a round bar that twists when it is loaded. It has the advantage that its weight is carried fully on the chassis; it can also be fitted in positions where height is limited. The disadvantage is that the amount of suspension travel which is allowed is very small.

Figure 10.6 Component layout on scooter.

Shock absorbers

The purpose of the shock absorber is to dampen the spring action and reaction. The shock absorber stops the motorcycle from bouncing each time it hits a bump in the road. There are two types of shock absorbers – these are telescopic and lever arm – and they are easily identified by their shape and mountings. On vintage machines friction dampers were used. These are formed by tightening two moving plates together. The suspension movement has to overcome the friction between the two parts. Between the two parts is a large brass watcher to prevent wear.

Telescopic shock absorber/damper

The telescopic damper/shock absorber gets its name from its telescope-like shape and action. The cylinder is filled with a special type of oil called shock absorber fluid. The lower part of the cylinder is connected to the suspension with a mounting eye. The upper part of the shock absorber, which comprises the piston and valve assembly, is attached to the vehicle's chassis with the upper mounting eye. The piston and the valve assembly move up and down inside the cylinder with the movement of the suspension

When the wheel hits a bump, the suspension travels upwards, shortening the distance between the mountings. In this situation the piston travels down the cylinder. The resistance of the fluid slows the movement of the piston, thus dampening the shock load on the suspension, when the wheel has traveled over the road bump and the suspension rebounds. That is, the wheel travels down again and the piston moves up in the cylinder as the distance increases again. The resistance of the fluid dampens the suspension movement and prevents the suspension from traveling too far. The shock absorber therefore dampens the suspension movement and stops the vehicle from bouncing like a rubber ball every time it hits a bump.

> **NOMENCLATURE**
>
> Bump is when the suspension is compressed; that is, the wheel goes up into the frame and the body goes down towards the road. Rebound is the opposite of bump; that is, the wheel goes down and the body goes up. Commentators on off-road races often use the word "jounce"; this is another word for bump – you'll see it when the bike lands back on the road after going over a humpback bridge. This take-off and jounce situation is referred to as "yumping," the Scandinavian pronunciation of jumping.

Bounce test

When you are checking the shock absorbers on a motorcycle, you should:

Figure 10.7 Lightweight wheel and single front disc.

- Inspect the mounting for damage or wear – the rubber bushes should be tight.
- Inspect the body for damage or dents.
- Check the seals for fluid leaks.
- Carry out a bump test – that is press down on the seat and let it go quickly. It should not go up and down more than three times

Wheel bearings

Hubs usually run on pairs of ball bearings. Larger, heavier motorcycles may use taper roller bearings. The bearings may be either preloaded, or they may be adjustable

Checking wheel bearings

With the bike supported on paddock stands, there are two checks to be made. One is for bearing free-play; the other for noise. To check for free-play, hold the top of the wheel with one hand and the bottom with your other hand. Try to move the wheel from side to side. If it moves more than the smallest amount, then the bearing needs either adjusting or replacing. To check for noise, spin the wheel by hand and listen; a good bearing should spin freely and quietly. Also look for oil/grease leaking from the hub seals – it is most likely bearing wear that allows excess travel and causes the seal to leak.

166 Motorcycle Engineering

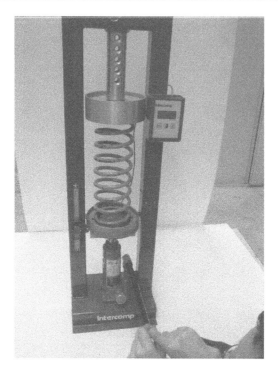

Figure 10.8 Checking spring rate.

Checking steering and suspension joints

Two of the main points of wear in the steering system are the ball joints and the track rod ends. To check these joints for wear, the two parts that the ball joints connect should be pulled in a direction that you would expect them to part. Using hand pressure, maybe with a small lever, a joint in good condition should show no signs of free-play at all.

WHEELS AND TIRES

> **Key points**
> - The wheels and the tires must be perfectly round, rigid, and correctly balanced.
> - There are rules relating to wheel and tire fitment and maintenance that must be followed; for example, the minimum tread depth on road bikes is 1.6 mm.
> - Wheels may be made from steel or aluminum alloy, with either a well base or detachable flanges for fitting the tire.

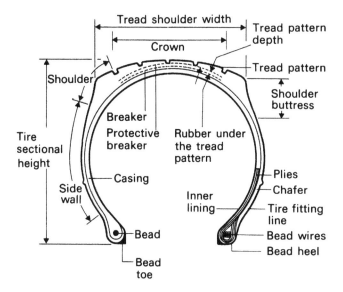

Parts of the tire

Figure 10.9 Parts of a tire.

- The tire and the wheel diameter are measured where the tire fits the rim; the width is measured between the flanges.
- Different types of tire construction and tread designs are used for different purposes.
- Specialist machines are used for changing tires and balancing the wheel and tire.

Functions of wheels and tires

The wheels and tires have a number of jobs to perform, namely to:

1. Allow the motorcycle to freely roll along the road
2. Support the weight of the machine
3. Act as a first step for of the suspension
4. Transmit to the road surface the driving, the braking, and the steering forces

Requirements of wheels and tires

For the wheels and the tires to be able to carry out their functions efficiently, they must be made and maintained to the following basic requirements:

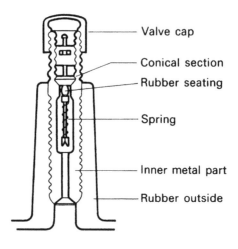

Figure 10.10 Schrader valve.

1. They must be perfectly round so that they roll smoothly.
2. They must be balanced so that the steering does not shake.
3. They must be stiff to give responsive steering and smooth running.

WHEELS

There are many different types of wheels in use. We'll look at the most common ones, that is steel well based, **aluminum alloy**, **wire-spoked**, and **two-piece**

Basic construction and sizes

Basically all wheels comprise a **rim** and a **wheel center**, which are attached together in some way. The rim is the part to which the tire is fitted.

The rim has a **flange** to hold the tire in place, a seating part to seal the tire bead against and retain the air, and a well section so that the tire can be fitted and removed from the rim.

The wheel center is the part that is attached to the hub. The wheel center usually has four holes for the **wheel studs**. The back of the wheel center has a flat section to make contact with the hub.

The **wheel diameter**, which is also the tire size, is measured at the tire seating part of the rim. You should note that the flange extends beyond this part of the wheel. The **wheel width**, which is also the equivalent of the nominal tire width, is measured between the inside faces of the flanges.

> **Tech Note**
>
> When fitting wheels, it is very important that they are tightened correctly. If they are not tight enough, they may come loose; if they are too tight, it can cause damage. To set the tightness correctly, you must use a torque wrench. Check the workshop manual, data book, or wall chart for the torque setting for each specific bike and type of wheel.

Aluminum alloy wheels

Aluminum alloy wheels were originally developed for aircraft; they give a combination of extreme lightness and high stiffness. The aluminum is alloyed with silicone for wear resistance, copper for hardness, and magnesium for easy casting. Aluminum alloy wheels are usually cast in one piece, but some specialist wheels are made in two parts that are held together with a ring of bolts. This construction allows the rim to split into two parts for easy fitting of the tire. Two-piece alloy rims are flat across the whole section; that is, they do not have a well section, as it is not needed.

The disadvantages of alloy wheels, as they are referred to, is that they are expensive, brittle, and have a limited life span. The brittleness is a problem if the wheel hits a curb or other hard object (called curbing); in this case the rim is likely to chip or crack, and obviously the tire will soon deflate.

Of course, alloy wheels can be made in a variety of styles, so they are available with looks to suit the bike and its owner. Good looks are in the eye of the beholder.

Alloy wheels are now being superseded by carbon fiber ones that are even lighter.

Wire spoke

The wire-spoked wheel is used on a number of classic bikes. The wire-spoked wheel is slightly flexible and springy. Spoked wheels, as they are usually abbreviated, have the advantages of being light, good looking, and allowing the air to pass through to cool the brakes.

The disadvantages are that they are easily buckled, especially if the spokes become loose; they are difficult to clean; and they need an inner tube to seal in the air.

TIRES

Construction

The basic construction of all vehicle tires is similar. The main components are the **tread**, the **casing**, the **wall**, and the **bead**. The wire bead forms the

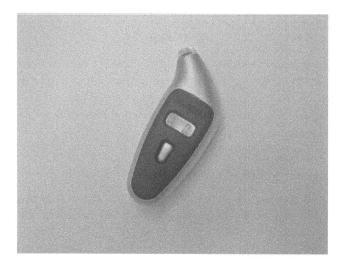

Figure 10.11 Electronic tire pressure gauge.

shape and the size of the tire. The textile plies and the rubber covering runs from one bead to the other. The two main types of tire construction are radial ply and diagonal ply.

Radial and diagonal ply

Radial ply tires are so called because the plies run in a radial manner. The plies of **diagonal ply** tires usually run at an angle of about 45 degrees to the tire radius.

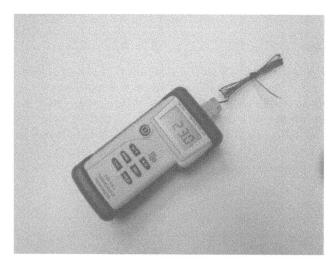

Figure 10.12 Digital thermometer.

Running gear and braking 171

Figure 10.13 Durometer to measure tire hardness.

Radial ply tires roll more freely than diagonal ply tires – this gives both better fuel economy and longer tread life. However, diagonal ply tires are quieter and less inclined to make the suspension or the steering knock or vibrate if a road bump is hit. For this reason, diagonal ply tires are used on

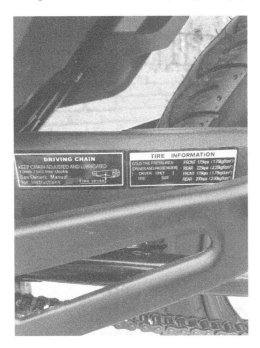

Figure 10.14 Tire pressure information sticker – found on most bikes.

some American vehicles – those likely to be used on poorly surfaced roads or off-road.

Because of their constructional differences radial ply and diagonal ply tires behave differently; for this reason, they must not be mixed on the same axle. If fitted, the radial ply tires must be fitted to the rear wheels. Motorcycle tires are usually supplied in matched pairs front and rear, with the rear being much wider on many machines. With tricycles, whether it is two front wheels or two rear wheels, the rule applies.

> **NOMENCLATURE**
>
> The term diagonal ply refers the fact that the plies run from bead to bead in a diagonal or angular manner. Previously this type of construction was more commonly referred to as cross-ply; this name is still used by older mechanics and tire fitters.

Tire treads

Different types of tire treads are used for different purposes. The tire tread patterns are designed for the different applications of the vehicle.

The purpose of the tire tread is to discharge water and enhance the grip. On average tires the tread is about 7 mm (1/4 in) deep when new. Off-road tires need to grip into the soft surface, so the tread is almost double the depth of a typical tire. High-speed tires have treads designed to give the maximum grip and run very quiet at high speed with minimum heat generation. Race tires usually use a much softer rubber compound for the tread to give better grip – this means that the tread will not last as long as for road tires.

Tire sizes

The size of the tire depends on the size of the wheel. The diameter of the tire is measured across the bead; it is the same as the diameter of the wheel rim. Most tire diameters are given in inches (in); increases in size are in 1-inch increments. A few specialist tires are measured in millimeters (mm) – called metric tires; these sizes always correspond to half-inch sizes between the inch increments of regular tires. The reason for this is so metric tires cannot be fitted to regular rims and vice versa.

The tire width on radial ply tires is given in millimeters; on diagonal ply tires it is given in inches.

Other markings

Between the width and the diameter numbers you will find a double figure number and a pair of letters. The double figure number is the aspect ratio of the tire, that is the height expressed as a percentage of the width. You will find aspect ratios between about 50% and 80%. Two typical letters that you will find are SR; the S shows that the tire is suitable for speeds up to 180 kph (112 mph); the R shows that it is a radial ply tire.

Tire pressure

The tire pressures should be checked regularly. Tire pressures must be kept within the limits specified by the manufacturer. In either hot or cold weather, the tire pressure may vary. The tire pressures also vary with the use of the motorcycle. The best time to check the tire pressures is before you set off in the morning. You should not alter them in the middle of a journey or event unless it is essential for a specific reason.

The use of an accurate tire gauge is essential; the pencil-type gauges are usually very accurate; for competition use, digital gauges give the best consistency of readings. Calibrated and certificated gauges are available for use with race and high-performance vehicles.

Tech Note

Ideal tire pressures are usually established during practice from lap times and alignment of the cross-tread temperatures.

Tire tread depth

The tread depth is set by European Regulations; currently the minimum depth allowed is 1.6 mm. This may be changed at any time

Because of the inconvenience of changing a wheel after a puncture and the risk of an accident or indeed closing a traffic lane, you are advised to regularly check all the wheels and tires, and if a tire fails, fit a new one.

Tire wear

If the tire pressures are set incorrectly, the tires will wear unevenly. If the pressure is too high, the tread will wear in the middle; if the pressure is too low, it will wear on the outer edges. Faults in the steering and suspension can also cause uneven tire wear.

Inner tubes

These are circular hollow rubber rings with a valve. The inner tube is used on motorcycles fitted with wire-spoked wheels to prevent air leakage. The inner tube is inflated inside the wheel and tire assembly. With wire-spoked wheels, a rim tape is fitted between the inner tube and the rim to prevent puncturing by the sharp edges of the spokes. In an emergency, punctures in inner tubes can be repaired with patches – these are larger versions of bicycle patches. However, this practice is not advised and should not be carried out on high-speed vehicles.

Tubeless tires

Most motorcycles use tubeless tires. That is, they do not have inner tubes; instead, the tire bead forms an airtight seal against the rim. Punctures in tubeless tires can be repaired by inserting a special rubber plug into the hole. This must not be done on the tire sidewall. High-speed tires should only be plugged as a temporary repair and the vehicle driven at a reduced speed.

Tire valves

Tire valves hold the air in the inner tube, or they are fitted to the rim in tubeless setups. The most popular type is the **Schrader valve**. When air is pumped into the tire, the valve core is forced downwards to allow the air to pass. When the intake of air is stopped, the pressure of the air in the tire helps keep the valve closed.

Wheel balancing

It is essential to balance the wheels and the tires as a unit to prevent wheel shimmy. Unbalanced wheels may cause the road wheels to shake from side to side, which will cause the steering wheel to shake from side to side. Wheel shimmy on out-of-balance wheels is usually noticeable at between 30 and 40 mph. Wheel balancing is carried out using a special machine; you will need a short training session for any specific machine.

Figure 10.15 Dual disc radial brakes and upside-down forks.

Running gear and braking 175

Figure 10.16 Heavy-duty rear adjustable shock absorber.

Tire fitting regulations

The laws on tire fitting and usage in the UK and most of Europe can be summarized as follows:

1. Radial or diagonal ply tires can be fitted to all vehicles.
2. If only two radial ply tires are fitted, these must be fitted on the rear wheels.
3. Radial ply and diagonal ply tires must not be mixed on the same axle.

Figure 10.17 Crash bungs on fork ends and carbon fiber mud guards.

4. The tire pressures must be kept within the manufacturer's recommendations.
5. The tread must be not less than 1.6 mm deep for the entire circumference over all the tread width.
6. The tread and the sidewalls must be free from large cuts, abrasions, or bubbles.

BRAKING SYSTEM

> **Key points**
> - Brakes on road vehicles must comply with MOT regulations.
> - Brakes work on the principle of converting kinetic energy into heat energy.
> - Disc brakes are less prone to brake fade.
> - A system of compensation is needed for even braking.
> - Hydraulic systems use the principles of Pascal's law.
> - Special care must be taken with brake dust and brake fluid.

The main function of the braking system is to stop the vehicle. The braking system also has two less obvious functions; these are to be able to control the speed of the vehicle gradually and gently and to hold the vehicle when parked on a hill or other incline.

The Highway Code gives a guide to typical stopping distances from different speeds. Road and weather conditions affect stopping distances greatly – in wet weather, it may take twice as long to stop as in dry weather.

The generally accepted formula is:

Stopping distance = Thinking distance + Braking distance + one vehicle length

It is important to remember that the braking distance increases with the square of the velocity (V^2). This means that it takes nearly twice as far to stop at 70 mph as it does at 50 mph.

MOT Requirements

To comply with the requirements of the **Driver and Vehicle Standards Agency** for motorcycles, scooters, mopeds, and motorcycle combinations (Class 1 and 2), the braking system must fulfill these basic minimum requirements:

- The brake lever and pedal must be in good working order, without stiffness or excessive travel.
- One braking system must record 25% efficiency, the other 30% efficiency.
- All the cables, pipes, and other components must be free from rust and/or damage and leaks.

(Full UK MOT details can be found in the MOT Testers Manual on the government website.)

> **NOMENCLATURE**
>
> DVSA is the current name of the part of the UK government agency that administers the policy on operating vehicles. Other countries have similar bodies, and their names change from time to time – people often just say department of transport. The term MOT – meaning Ministry of Transport (now defunct) – is still used on official documents. In the United States, the equivalent department is referred to as DOT.

Friction

Friction is the resistance of one body to slide over another body. It is only dependent on the surface finishes of the materials; size of contact area does not affect the friction. When two areas are in contact and force is applied to hold them together, the friction generates heat.

The braking system uses friction to convert the **kinetic energy** of the vehicle into **heat energy,** which is dissipated to the atmosphere.

> **NOMENCLATURE**
>
> Dissipate is a scientific word for dispelling, getting rid of, or spreading about. The heat from the brake pads and discs is dissipated into the atmosphere so that it does not build up and allow the brakes to get hot.

The amount of **heat generated** by the brakes to stop a motorcycle from, say, 200 mph (320 kph) will be almost the same as that generated in the engine to accelerate the car to the same speed.

The amount of friction depends on the materials of the **friction surfaces,** that is, the **pads** and the **discs,** or the **drums** and **shoes.**

The heavier and faster the motorcycle, the bigger the brake components will need to be to dissipate the greater amounts of heat generated.

> **Tech Note**
>
> Kinetic energy is the energy of motion – the faster a vehicle is traveling and the heavier it is, the more kinetic energy it possesses – so the hotter the brakes will get on a twisty circuit or a downhill section where the brakes are in constant use.

MECHANICAL BRAKES

> **SAFETY NOTE**
>
> Brake dust may contain asbestos; in any case it can cause breathing problems, so use a breathing mask when appropriate and clean down brake assemblies with a proprietary product.

A rod or a cable is used to transmit the effort from the lever, or pedal, to the brake shoes. The mechanism used to move the shoes is usually a simple cam arrangement. When the cable, or rod, is pulled the cam is turned so that the brake shoe is pressed against the drum.

Mechanical brakes are very simple, but they are subject to a number of problems:

- Cable stretch
- Wear of the connecting clevis pins and yokes
- Need to adjust each cable run separately

Figure 10.18 Brake pedal.

Running gear and braking 179

Figure 10.19 Brake return spring.

Figure 10.20 Drum brake on rear of lightweight 50 cc. You can see the adjustment nut.

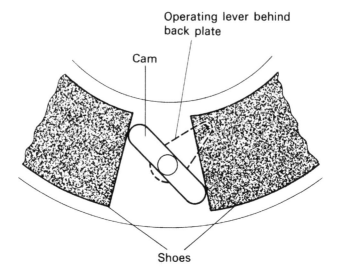

Operating cam on mechanical brakes

Figure 10.21 Operating cam on mechanical brakes like those in Figure 10.20.

Compensator

Because it is essential to apply equal stopping force to each wheel on an axle, with tricycles some form of compensation device is used in the system. This is a mechanical device that distributes the force evenly between the brake assemblies (drum or disc) so that the vehicle will stop evenly without skidding, or it holds the hand brake evenly on both wheels.

There are two main types of compensators: swinging link, also called swivel tree, and balance bar.

The swinging link has three arms and is mounted on the axle casing. The longitudinal cable pulls on the longer arm; the shorter arms pull at right angles to the longer one that is turning the force through 90 degrees. This changes the direction to transverse and gives a mechanical advantage (leverage) by the difference in arm lengths. It is important to check that the mechanism moves freely – old systems may have a grease nipple on the swinging link to encourage lubrication.

The balance beam is used on most current hand brakes, forming the lower part of the lever. The center of the beam, which is a very short metal component, is free to twist in the hand brake; the two cables, one to each wheel, are attached to each side of the beam. This allows for slight variations in cable length caused by cable stretch or uneven adjustment between the two brake assemblies.

HYDRAULIC BRAKES

There are many types of hydraulic brake systems and variations within those systems. The advantage of hydraulic brakes is the hydraulic system is self-compensating – there is no need for mechanical compensation systems and the attendant adjustments to be made; this is explained by Pascal's law.

Pascal's Law

Pascal was a French scientist who lived between 1623 and 1662; he spent a lot of time studying why wine bottles broke at the bottom when the cork was pushed in the neck. His discovery, that a liquid cannot be compressed and that any pressure applied to a fluid in one direction is transmitted equally in all directions, led to the invention of hydraulic brakes about 300 years later.

Let's have a look at this in more detail. Pascal's theory is applied so that when the rider presses the brake pedal, a much larger pressure is transmitted to the brake pads and shoes.

Pressure = Force / Cross – sectional area

Force = Pressure * Cross – sectional area

For example, if the rider presses on the brake pedal so that a force of 400 N is applied to the master cylinder, which has a cross-sectional area of 50 mm², what force will the wheel cylinder piston exert on the brake shoes if its cross-sectional area is 100 mm²?

$$\text{Pressure at master cylinder} = \text{Force / Cross-sectional area}$$
$$= 400 / 50$$
$$= 8 \, N/mm^2$$

Pressure at wheel cylinder = Force/Cross-sectional area

Applying Pascal's law, the pressure at the wheel cylinder is the same as that at the master cylinder, but the area is different, so we can insert the numbers that we know; therefore:

8 N/mm² = Force/100 mm²

Transposing the formula:

Force = 8*100

Force = 800 N

Figure 10.22 Brake fluid reservoir.

As you can see, doubling the cross-sectional area doubles the force. So, if the brakes were made using four-wheel cylinders or disc brake pistons, each of which is double the cross-sectional area of the master cylinder, then the force applied by the driver will be multiplied by 8 at the brakes.

$$\text{Pressure} = \frac{\text{Force}}{\text{Cross-sectional area}}$$

the difference in cross-sectional areas. Piston B is twice the area of piston A, so the force on piston B is twice as great.

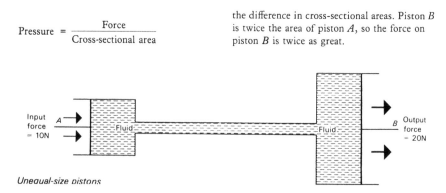

Uneaual-size pistons

Figure 10.23 Principle of hydraulic brake system.

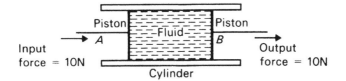

Equal-size pistons in a cylinder

Figure 10.24 Equal-size pistons.

Layout of simple system

The simple system on motorcycles is to have two independent braking systems: one hand-operated lever and one foot-operated lever. The hand-operated lever is usually the right hand, as this is usually the strongest hand and for most people the most responsive. On older bikes the foot lever is operated by the left foot; the gear pedal was on the right. Since the change in engine and gearbox layouts, the brake pedal is now on the right and the gear pedal on the left.

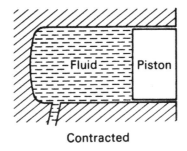

Contracted

Wheel cylinder (single piston): contracted

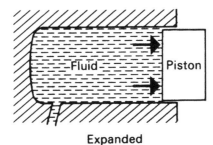

Expanded

Wheel cylinder (single piston): expanded

Figure 10.25 Operation of wheel cylinder.

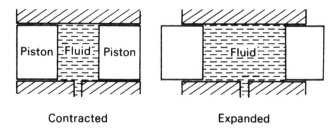

Brake wheel cylinder (double piston)

Figure 10.26 Double-piston wheel cylinder.

> **SAFETY NOTE**
>
> The pressure of the brake fluid with the pedal depressed is about 50 bar (750 psi), which is higher than that of the air from the compressor – and you know that must be handled with care – so make sure that the pressure is released before working on the hydraulic system.

DRUM BRAKES

Vintage bikes use drum brakes front and rear; currently, drum brakes are only found on the rear of small bikes and mopeds.

Drum brake shoe layouts

Current vehicle rear drum brakes use **leading and trailing brake shoes**; older race vehicles fitted with drum brakes at the front usually have **twin leading shoes** at the front.

The twin leading shoes give the maximum braking power – the leading edge of the shoe tends to dig into the drum and give what is referred to as a self-servo action. When stopping hard the front brakes do the bulk of the work – typically the front brakes do 70% of the work compared to 30% at the rear. However, twin leading shoe brakes are only efficient when the vehicle is going forwards, so they are not useable for a hand brake where the vehicle may be parked on a hill pointing either up or down. So, for hand-brake operation a leading and trailing shoe arrangement is needed.

Wheel cylinder

There are two main types of wheel cylinders:

- Single piston – used on twin leading shoe brakes where one cylinder operates each shoe

- Twin piston – used on leading and trailing brakes so that the wheel may be held in either direction of rotation; this is needed to prevent rolling back on inclines

> **NOMENCLATURE**
>
> Pipes, hoses, Bundy, and lines are all terms that are used and misused in the motorcycle industry. The same applies to couplings, fitting, connectors, brake nuts, and pipe ends. You should try to use the technically correct terms always. However, when building a braking system on a bike, you will find reference in the workshop more to the manufacturer or type, such as Goodridge, AP, and Bembo.

Brake Shoes

The brake linings are attached to the brake shoes by either rivets or a bonding process (glue). The shoes on ordinary road vehicles are usually fabricated (welded) from plain steel; on race vehicles they are made from aluminum alloy for lighter weight and better heat dissipation. Usually the brake shoes are held in place against the wheel cylinders by strong springs.

> **Tech Note**
>
> Before stripping drum brakes, it is prudent to either sketch or photograph the layout of springs on shoes.

> **SAFETY NOTE**
>
> Brake shoe springs are very strong. Do not put your fingers where they might get trapped or hurt by the springs slipping.

At every major service, the brake linings should be checked for wear. The workshop manual will give a minimum thickness figure – typically 3 mm (1/8th inch). If the linings are riveted in place, the rivet heads must be well below the surface of the brake shoe.

Drums

Brake drums may be made from cast iron or aluminum alloy with a steel insert for the friction surface. The aluminum alloy ones are much lighter than the cast iron ones – possibly only one-third of the weight. They may also have cooling fins cast on them to help prevent brake fade.

The brake drums are usually made as part of the hub. When inspecting the brake pads, the drums should be inspected as well – look for score marks, cracks, and ovality.

> **NOMENCLATURE**
>
> Ovality means oval, or in the setting of brake servicing – not properly round. This problem causes brake grab.

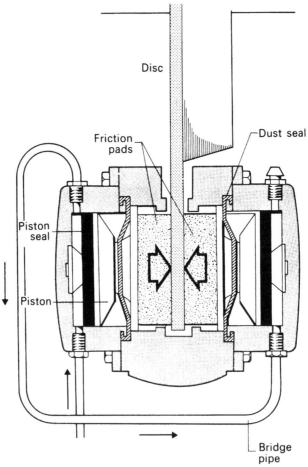

Disc brake caliper

Figure 10.27 Disc brake caliper with two pistons.

Running gear and braking 187

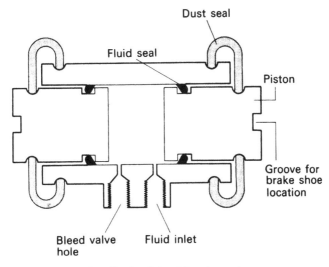

Brake wheel cylinder (double piston)

Figure 10.28 Double-piston wheel cylinder.

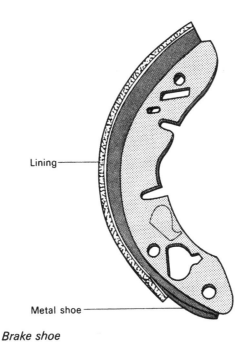

Brake shoe

Figure 10.29 Brake shoe.

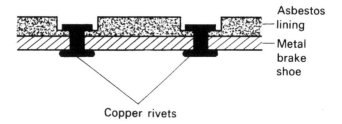

Cross-section of brake shoe

Figure 10.30 Riveted brake lining showing rivets below lining surface. Asbestos is actually a synthetic material.

Disc brakes

Disc brakes have several advantages over drum brakes, namely:

- Less susceptible to brake fade – that is a reduction in braking efficiency through an increase in the temperature of the friction surfaces, usually after several successive brake applications
- Open to the air and therefore kept running cooler
- Easy-to-change pads
- Greater braking effort for size and weight with the aid of a brake servo
- Self-adjusting

Tech Note

Jaguar's car racing reputation was built on winning the Le Mans 24 Hour Race many years ago. They did this by using Dunlop disc brakes to enable them to brake later and harder into each corner, thus increasing their lead on each lap. It was 20 years later before they appeared on motorcycles.

Calipers

Calipers are the disc brake equivalent of the drum brake wheel cylinders. The fluid moves the pistons to press the pads against the disc. There are many variations of calipers, from single-piston calipers to ones with six pistons used on race and high-performance motorcycles.

NOMENCLATURE

You will hear terms such as **four-pot** calipers: this means that each caliper has four pistons – two on each side.

As the pads wear, the pistons will expand out to take up the wear. To return the pistons in some cases it is necessary to use a special tool to turn the piston while pushing it back into the caliper – look in the workshop manual for the specific application.

Discs (U.S. rotors)

The pads act on the discs (rotors) to give the necessary friction. Plain discs are simply cast iron, but most discs are coated with some form of surface finish to improve braking and resist corrosion. The discs are bolted to the hubs.

When overheated, discs can warp like a buckled wheel on a bicycle. This warping will give uneven braking and can often be felt at the brake pedal. When replacing pads, it is a good idea to check the discs for warping, or as it is called, run-out.

Special discs

Discs may be manufactured in a number of different ways, mainly to improve cooling. Some examples are:

- **Vented discs** – a gap between the two sides of the disc with air vents fitted
- **Cross-drilled discs** – as its name says, drilling across the disc
- **Wavy discs** – a wavy edge to improve braking effort give a number of leading edges

On very advanced race and high-performance motorcycles **carbon discs** are used. These are usually used as multifloating discs – a system used on large aircraft.

> **SAFETY NOTE**
>
> When replacing brake pads, do one side at a time, and do not operate the lever or pedal to ensure that the seals are not broken and no fluid is lost.

Brake pad thickness

As the brakes are used the **friction material** wears, and there is a point at which it becomes necessary to replace the pads. This is when the thickness of the friction material reaches a minimum amount. The vehicle manual will state this figure and how to **measure** it.

Brake pad grade

A variety of different materials are used for brake pads. Basically these fit into two major classifications:

- **Soft pads** give more friction, but generally, they create more heat and are therefore prone to fade.
- **Hard pads** give less friction, less heat, and last longer. Hard pads are used for race bikes.

Brake lines

Brake lines connect the brake components, transmitting the fluid pressure according to Pascal's law with a small fluid displacement. A number of different materials are used for brake lines. On road bikes, they tend to be an alloy steel; on historic racers, copper is frequently used to look authentic and for its ease of bending; off-roaders tend to use stainless steel for its high value of hoop stress – that is, its resistance to expand under pressure. Stainless steel braided hoses are available as after-market fitments for most motorcycles.

Brake fluid

Brake fluid is a special type of oil developed to give the specific properties needed by the braking system. These are:

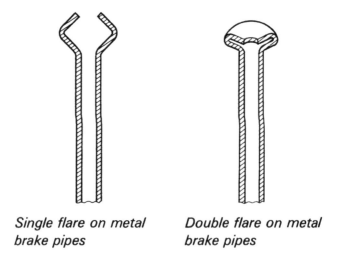

Single flare on metal brake pipes   *Double flare on metal brake pipes*

Figure 10.31 Single-flare and double-flare brake pipe ends.

Running gear and braking 191

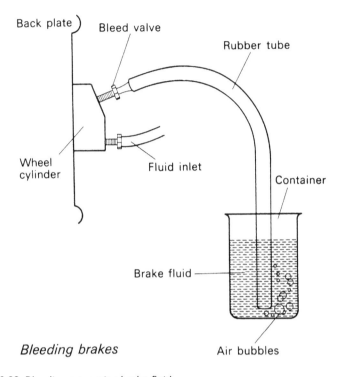

Figure 10.32 Bleeding or venting brake fluid.

- High boiling point to reduce the risk of brake fade
- Noncorrosive to rubber and the other materials used in the braking system
- Lubricating properties for the mating parts
- Will not fail under very high pressure
- Low viscosity for rapid response to pedal operation

When topping up or replacing brake fluid, ALWAYS USE THE RECOMMENDED BRAKE FLUID; BRAKE FLUIDS MUST NOT BE MIXED.

SAFETY NOTE

Remember brake fluid removes paint – be careful when you are using it.

Tech Note

Hose spanners are available to reduce the risk of damaging brake hose ends.

Brake adjustment

When the friction surfaces wear, adjustment is needed to prevent excessive brake pedal travel. There are two main types of adjusters: **wedge** type and **snail cam** type

---
Tech Note

What are self-adjusting brakes?

There are two main types of self-adjusting brakes: disc brakes are self-adjusting by piston movement in the calipers; rear drum brakes are self-adjusting by using a ratchet mechanism on the hand-brake mechanism inside the drum.

Fluid venting

This is also referred to as **bleeding the brakes**. That is the process of **removing air** from the braking system after repairing or replacing one of the hydraulic components. There are two main ways of doing this:

- **Manually** – A rubber tube is attached to the **vent nipple** (bleed nipple) by pushing it over the nipple to give a firm fit. The other end of the tube is submerged in a container of clean brake fluid. The nipple is slackened about one turn. When the brake pedal or lever is depressed, fluid will flow and bring with it any trapped air. The master cylinder reservoir must be kept topped up throughout this job. When air bubbles stop flowing, the nipple is retightened and the pedal tested for a firm feel.
- **Using air pressure** – A device is attached to the master cylinder reservoir, which supplies fluid under pressure using air pressure from either a hand pump or a compressed air supply. The nipple is attached with a tube in the same way as in the manual method. There is no need for anybody to press the pedal. When the nipple is opened fluid will flow – simply close the nipple when air bubbles stop.

Tech Note

Brake fluid is hygroscopic; in other words, it attracts water from the atmosphere – it should be replaced at least every couple of years to keep the system fresh.

Antilock braking systems (ABS)

Unlike with a car, where a locked wheel will just cause a skid; a motorcycle locked wheel usually leads to both the bike and rider hitting the road.

Figure 10.33 Radial brake mounting incorporated into fork end.

Honda first devised a system where the front and rear brakes were combined with a balance valve mechanism to divide the force between them. This saved the rider having to "feel" the braking on each wheel. Good riders could feel, or sense, braking limits. The current ABS systems do all that now.

Each wheel has a sensor which tells the control unit the relative velocities. This sends a signal to the control unit when brake lock is imminent to reduce the braking force on that wheel. It does this in a tenth of a millisecond – 0.0001 second – faster than the brain can react.

Radial brakes

The front brake calipers are attached by bolts to the bottom of the forks. Conventionally the bolts went in from the side. The current trend is to fit the bolts radially, that is, towards the axle.

Figure 10.34 Rear wheel mounting showing chain adjustment nut and marks on swinging arm.

SOMETHING TO THINK ABOUT

You are traveling along a German Autobahn with your partner on the pillion. Travelling steadily at 40 m/s (90 mph). You have good brakes giving 75% efficiency. Suddenly you need to stop. How long will it take you?

Thinking distance = 40m / s * 0.25s
= 10m

Deceleration = 75 / 100 * 9.8 (gravity)
= 7.35 m / s^2

Formula for braking $V^2 = U^2 + 2AS$
V is final velocity zero, U is initial velocity 40 m/s

$0^2 = 40^2 + 2 * (7.35) * s$

s is the braking distance
s = 1600/14.7
s = 109 m

Shortest possible stopping distance = thinking distance + braking distance

= 10m + 109m = 119m

One hundred and nineteen meters is longer than most international football pitches.

Figure 10.35 Radial disc setup on Zero Electric Motorcycle.

Running gear and braking 195

Figure 10.36 Radial brake lever on Zero Electric Motorcycle.

Chapter 11

Motorcycle electrical and electronic systems

> **Key points**
>
> - The battery is the central store of electricity for the vehicle.
> - Battery electrolyte is very corrosive – avoid splashing it on your skin or the vehicle.
> - The alternator is driven by the crankshaft to keep the battery charged – not all motorcycles have them.
> - Many motorcycles use the metal frame as earth return.
> - Race bikes often use a separate trolley battery for starting.
> - Vehicles may have several ECUs – engine on road/engine on track/body controls/transmission controls and other separate systems.

> **SAFETY NOTE**
>
> Electrical short circuits are the main cause of vehicle fires. When working on an electrical system, it is easy to cause a temporary short circuit by two terminals or cables touching. For this reason you MUST always DISCONNECT THE BATTERY before carrying out any work on the electrical circuit.

BATTERY

The battery is the central store of electrical energy for any vehicle. Its purpose is to store electrical energy in chemical form. Most vehicle batteries operate at a nominal 12 volt. Batteries are also available as 6 volt and 24 volt. To control weight distribution, or give added power, or reliability, vehicles may use more than one battery. Two 6-volt batteries may be connected in series to give a 12-volt supply; two 12-volt batteries may be

Figure 11.1 Battery being filled with electrolyte.

connected to give a 24-volt supply. A 12-volt battery may have three terminals to allow both 6-volt and 12-volt supplies.

NOMENCLATURE

The word battery means a group of things and came into use in the seventeenth century as a place where guns and soldiers of the English civil war were located. In Hampshire there is a village called *Oliver's Battery*. The technically correct name for a car battery is an *accumulator*; you may come across this term when you are working at an advanced level with your studies.

Casing – This is made from a nonconductive material that can withstand low-level shock loads; inside it is divided into compartments, or cells.

Cells – Each cell is made up of positive plates and negative plates divided by porous separators and submerged in electrolyte. There is always an odd number of plates – there is one more negative plate than positive plates to make the maximum use of the positive plates.

The nominal voltage of each cell is 2 volts, so six cells are needed to make up a 12-volt battery. We say nominal voltage, as it is not the exact voltage. The voltage given by each cell varies with the state of charge of the cell, and therefore the battery. A fully charged cell will produce 2.2 volts, meaning that the fully charged battery will produce 13.2 volts.

Electrical and electronic systems 199

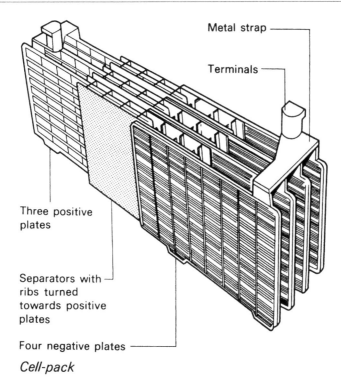

Figure 11.2 Battery cell pack. Each cell pack gives 2.2 volts when fully charged.

Electrolyte – A solution of sulfuric acid. It is highly corrosive – treat it with care. Batteries may come dry charged, in which case the electrolyte is added only when the battery is purchased. This aids storage and transportation. Or they may be wet charged – the battery is charged after the electrolyte is added.

The electrolyte in motorcycle vehicle batteries is often in a gel form to prevent spillage in the case of an-off.

SAFETY NOTE

Electrolyte, usually called **battery acid**, is very **highly corrosive**; it will burn the skin off your hands and holes in your overalls. Read the COSHH sheet and take extreme care – **treat batteries with respect**.

Battery charging – Batteries are kept fully charged when on the motorcycle by the alternator (or dynamo on older bikes). When not on the vehicle a battery charger is needed. It is wise to keep a battery fully charged. The battery off a race bike, for example, should be fully charged before it is

Table 11.1 Battery chemistry

	Fully Charged	Discharged (flat)
Positive plate	Lead peroxide (PbO_2)	Lead sulfate ($PbSO_4$)
Negative plate	Spongy lead (Pb)	Lead sulfate ($PbSO_4$)
Electrolyte	Strong sulfuric acid ($2H_2SO_4$)	Dilute (weak) sulfuric acid ($2H_2O$ with a percentage of $2H_2SO_4$)

stored for the off-season period. **Maintenance-free batteries** must only be charged at low amperage to avoid damage to the plates; check that the battery charger is the correct type for use with maintenance-free batteries. There are three types of battery charger in use; these are constant current charging, constant voltage charging, and taper charging.

Charging the battery alters the chemical structure in the plates and the electrolyte acidity.

Relative density (specific gravity)

> **NOMENCLATURE**
>
> Relative density (RD) is the technically correct term for specific gravity (SG); in motorcycle engineering they are both used to mean the same thing. Water at 4 °C (39 °F) has a weight of 1 kg per liter. RD and SG compare the weight of other liquids to this, so sulfuric acid is heavier than water.

The RD or SG of a battery electrolyte is an indication of the state of charge of a battery. This is checked using a hydrometer; this is a glass tube

Table 11.2 Relative density readings

Color	RD or SG	State of Charge	Comment
Green	1.280	Fully charged	Leave battery to cool after charging before testing
Yellow or orange	1.160	Half charged	
Red	1.040	Discharged (flat)	May vary with temperature

with a float inside it. The denser the liquid, the higher the float will be in the electrolyte. The float may have a numbered scale or simply a colored scale to indicate the state of charge.

Maintenance – The following points should be checked:

- **Electrolyte level – Top up with deionized water** as needed. Sealed maintenance-free batteries should never need this activity.
- **Keep batteries fully charged** – If the vehicle is laid up for the off-season, keep it charged with a bench charger.
- **Use a proprietary protector** (petroleum jelly or similar) on battery terminals to prevent corrosion.
- **Keep terminals and connectors clean and tightly fastened.**

About half of all breakdowns are caused by battery faults. The AA (Automobile Association) and the RAC (Royal Automobile Club) frequently publish statistics on causes of breakdowns.

Starter pack (or **power pack**) – For breakdown and recovery work, a special battery in the shape of a small briefcase is used. This is simply charged from the mains with its own cord and plug (230 volts UK; 110 volts Europe, United States, and most other countries).

Race battery – Batteries for race vehicles are very light and usually of the gel type. They do not need to start the vehicle, as a portable trolley battery is used for starting. That is a large heavy-duty battery mounted on a trolley. Or a starter pack system may be used, depending on the size of the engine.

RACER NOTE

Keep your trolley battery fully charged, and remember which side of the vehicle it plugs into.

ALTERNATOR

The alternator is driven by the crankshaft; it produces alternating current (AC), which is converted into direct current (DC) to charge the battery. The alternator is made up of a rotor, which spins inside a stator and is encased in adjoining front and rear casings. The electricity is generated by the movement of the rotor inside the magnetic stator. The electricity leaves the rotor through two slip rings and brushes. The rear casing houses the rectifier and the regulator. The rectifier converts the AC current into DC current; the regulator uses a set of diodes to control (regulate) the amount of current supplied to charge the battery. Too much electricity would cause the battery to boil and become useless; too little and the battery would become flat. Typically, it takes 20 minutes of running the engine to recharge it to replace the electricity used in starting the bike from cold. Much less electrical energy is used starting the engine from warm than starting from cold. Maintenance of the alternator is minimal; read the output using the engine diagnostic tester if a fault is suspected.

Dynamo – Before the alternator was the dynamo. This ran at much slower speeds because the commutator's soldered segments would break away above a specific speed – around 6000 rpm. Dynamos are usually mounted at the back of the engine above the gearbox on preunit designs. The dynamo is usually driven by a gear off the camshaft.

Magneto – This is used to generate HT (High Tension) for the ignition circuit on small-capacity motorcycles, where a power supply is also needed. The magneto is combined with a dynamo to make a **mag-dyno**. The mag-dyno is often used on sporting trials motorcycles. The power for the lighting (this is the only use of electricity apart from the ignition) is only provided when the engine is running, and is also dependent on engine speed.

> **NOMENCLATURE**
>
> Alternator, dynamo, and mag-dyno are all referred to as generators, as they all generate electricity.

STARTER MOTOR

There are two major types of starter motor:

- **One-way clutch** – This uses a clutch drive from the motor to the flywheel so that a mechanical linkage is made only when the starter is turning fast enough; when the engine starts, it is disengaged.
- **Pre-engaged type** – This has a solenoid to engage the gear on the starter motor with the flywheel ring gear before the starter motor turns.

The actual motor part is similar on both types. The **armature** is turned inside the **field coils** of the motor by the effect of supplying a large electrical current to the starter motor brushes. Because of the heavy current involved, a special switch called a **solenoid** is used. This is operated by power supplied to it when the key is turned to the start position. On bikes a separate starter button is used, the ignition being controlled by a key switch with on or off only.

As a guide a starter motor takes about 180 amps to turn an engine from scratch. Engines typically need to rotate at 120 rpm to initiate

Electrical and electronic systems 203

Figure 11.3 Starter motor on 50-cc engine.

combustion; some engines can achieve this in less than half a turn of the crankshaft.

> **SAFETY NOTE**
>
> If you are testing a starter motor off the vehicle, do so in a suitable rig – not loose on the bench top, as the current required could cause a burn or start an explosion.

VEHICLE CIRCUITS

Electrical symbols to make electrical circuit diagrams easy to understand and small enough to be printed on A4 paper. There are variations between countries and manufacturers, but it is usual for them to be easily identified intuitively even if you see them for the first time.

Circuit diagrams – Most motorcycle manufacturers provide circuit diagrams – the student should seek out those of bikes which are of interest.

Figure 11.4 Horn mounted on frame.

Figure 11.5 Switch to prevent riding off with the stand in the down position.

> **Tech Note**
>
> Beware when using circuit diagrams. They are like the London Underground map; the position of a component on the diagram is not an indication of the position of it on the vehicle.

Chassis earth – This takes large amounts of current, especially when starting the engine (typically 180 amps). It is essential to check all chassis earth points. There is a major lead between the battery and the chassis, and a similar lead between the chassis and the engine.

> **Tech Note**
>
> It is prudent to check all earth connections for tightness as part of your spanner check procedure – also use a suitable proprietary anticorrosion coating to keep them in good condition.

Figure 11.6 Dashboard with ignition key switch, tachometer, and speedometer.

Cables and connectors – All cables on motorcycles use crimped ends into semi-locking plastic connectors. The type used varies with the manufacturer. The student is advised to investigate those of the manufacturer which is of interest.

On older motorcycles mainly copper wire is used – as compared with the aluminum or gold-plated wires of current high-performance bikes. Copper wire is usually soldered to brass or tin plated connectors on older bikes. The soldering is necessary to give good electromechanical connection and prevent the wires from coming off the terminal.

Tech Note

Soldering is a good skill to learn if you intend to make up some of your own cable and connectors with high electrical and mechanical integrity.

Figure 11.7 Top red rocker switch is emergency kill button, center is light switch, lower red is starter button.

Electrical and electronic systems 207

Figure 11.8 From Figure 11.7, with brake fluid level indicator on the left.

Figure 11.9 Blue switch on top is headlamp dip switch, black one in the middle is indicator switch, yellow push button is horn.

Figure 11.10 Yellow button is for pass-light.

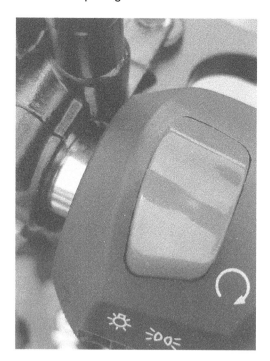

Figure 11.11 Red kill switch.

LIGHTS

Side lights are fitted to indicate the size and position of the vehicle to other motorists; in the UK they must be white in the front and red in the rear. Other countries use amber lights in the front. There are regulations as to the position on the vehicle and minimum size. Until recently a 5-watt lamp was compulsory; now light-emitting diode (LED) lights are permitted.

Headlights are fitted to give the rider clear vision in darkness. The law sets regulations on lamp position and beam placement – the rules are devised to prevent them from dazzling other motorists. Headlights may be a number of different colors, but white is normal.

Spot lights and fog lights – Used as their names imply. They should only be used in appropriate conditions.

Direction indicators must be amber in color and flash at a rate of 60 flashes per minute. Their position is also subject to regulations, and side repeaters may be used too.

Brake lights are operated by the brake pedal and brake lever. The minimum is 21-watt bulbs or equivalent LEDs.

Running lights – On U.S. vehicles and in some other countries too. These may be the equivalent of the side lights – usually amber in color.

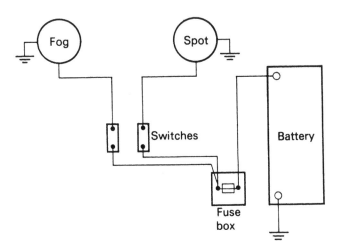

Spot and fog lamp wiring cirucit

Figure 11.12 Auxiliary light wiring circuit.

210 Motorcycle Engineering

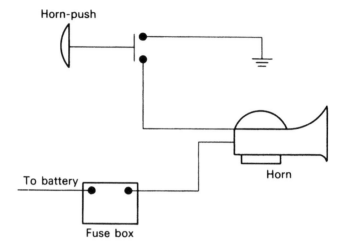

Horn circuit

Figure 11.13 Horn wiring circuit.

Figure 11.14 Portable tool kit.

Electrical and electronic systems 211

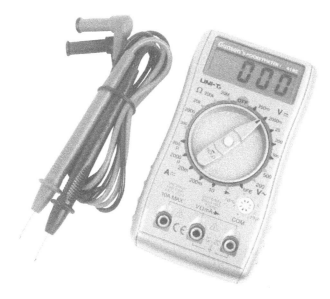

Figure 11.15 Pocket multimeter.

Figure 11.16 View of rear light cluster.

Chapter 12

The motorcycle industry

The motorcycle industry, like many other industries, is going through a large-scale change. But this is not new – the motorcycle is constantly being developed, and hence the ways of making them and selling them are constantly changing too. There is a saying that nothing is constant except change – there are likely to be disruptions to the motorcycle industry for as long as people ride motorcycles and others want to make a living from them. This chapter looks at the industry past, present, and the possible future.

> *I hope that you may read this chapter and gain an insight that you will use to start a new business, or to improve a current one, so as to keep the motorcycle industry alive for future generations to enjoy.*

TRADITIONAL MOTORCYCLE SHOP

Motorcycle retailing and repair in the form of the traditional motorcycle shop has been in this form since motorcycles came into common usage about 1900, with great growth in the years after World War II. The traditional shop will supply new motorcycles, deal in used ones, sell accessories and clothing, and carry out service maintenance and repair work. There are currently many traditional motorcycle shops in the UK, Continental Europe, and the United States that can trace their beginnings back 100 years or so. There are also many of the new style – rather like car dealerships.

The manufacturers of motorcycles have tried to change the traditional shop by introducing dealerships and closed supply chains. That is saying to the dealers you must only sell our make within this geographical area in the same way that has happened in car sales. However, this is changing more now that manufacturers are selling direct to the customer through their websites, thus cutting out the traditional shop completely. This often leads to problems if the motorcycle breaks down or needs other forms of service; the dealership may choose not to offer the customer what would be regarded as a normal level of service – why should they, as they made no money from selling it? When the author worked for the RAC, resolving such issues was a common problem. Whether buying or selling used motorcycles, the inclusion of an independent warranty and RAC/AA membership is advised in all cases.

Quality, old-school motorcycle shops will perform a service check all machines before sale and offer a comprehensive return-to-shop warranty. There's an old adage: you only get what you pay for – look for a shop that is run by motorcycle riders. The mail-order giants of the motorcycling world can afford to sell at low prices for two main reasons: one is volume of turnover; the other is that they make more money on arranging finance and insurance for the machine. It's worth looking at the possible profit streams available to the larger motorcycle shops:

- Mark-up on the machine – typically 30% of sale price
- Fee for arranging hire-purchase or other payment method – typically 5% of total
- Extended warranty – typically 25% of this fee
- Door-to-door delivery – typically 30% profit
- RAC/AA/other breakdown get-you-home schemes – typically 25% of this fee
- Insurance – typically 12% of the cost
- Security/antitheft devices – typically 40% of price
- Administration fees – vary up to about £100 ($100)
- Additional clothing and helmets and accessories – the amount of mark-up varies up to about 80%

Most traditional motorcycle shops will also offer MOT facilities and collection and delivery at cost.

How do you recognize a good motorcycle shop? There are a few giveaway signs in the author's view:

- A few oil marks on the floor – not oil slicks though – this shows usage.
- A chirpy "hello" or similar.
- A face that you've seen at a local motorcycle event or in *Motorcycle News*.
- A range of machines and prices.

Less than two of the four – walk out.

If you are selling anything, be prepared for the unexpected. A motorcycle salesman I know had a surprise one Saturday afternoon. A man turned up to the shop on an old bicycle and asked to look at the latest superbike. He was very keen, but the shop was busy. After a while, the bike was started

Figure 12.1 Underground personnel entrance to a major manufacturer's premises.

up and he sat on it. The salesman was getting a bit edgy. Half an hour later, a Range Rover pulls up at the door of the shop driven by a fashion model. The bicycle rider hands over his debit card, puts the old bicycle into the Range Rover, gets his helmet from his model wife, and waits to ride the superbike home.

Manufacturing

Total motorcycle sales in the UK average about 10,000 units per month. The three main categories are naked machines, scooters, and adventure sport machines – in that order. There is always a small number, usually single figures, of an "unspecified" category. Whether you are aiming to get into one of the top-selling groups or just make it in to the single figures, you need to look at what is involved in manufacturing. This also applies to the manufacture of accessories, custom parts, and tuning items.

Motorcycle manufacturing, like any other manufacturing, is about return on capital investment per unit made. The more units that can be made for a given capital investment will spread the cost further, allowing lower prices and greater profits. Let's briefly look at this.

If you are setting up manufacturing motorcycles or parts, you have three sets of costs to consider:

1. **Initial capital costs**
2. **Fixed running costs**
3. **Variable manufacturing costs**

This model also applies to retail organizations as well – in fact, almost any business.

Initial capital costs – The amount of funding needed to set up the business. This includes:

- The building – For manufacturing, this usually entails some form of specialist construction for delivery and loading areas, ventilation for the specialist processes such as welding and paint spraying, high level of electrical, and other power supplies.
- The tools and machinery – Jigs, welding equipment, bending and forming tools, paint/finishing equipment, drilling, and other machining equipment.
- Operating funds – The money needed to pay for initial materials, labor costs, and up-front charges such as rent, business rates, energy supplies, and advertising/marketing.
- Intellectual property – This can be the most valuable item: the designs and ideas for the motorcycle. A good example is the purchase of Harris Performance by Royal Enfield. The trade names, know-how, and intellectual property (IP) of Harris Performance immediately gave Royal Enfield a podium position as a motorcycle manufacturer.

One of the problems with the motorcycle industry is the lack of investment in manufacturing. In fact, this applies across all manufacturing. Manufacturing needs large investments that do not see immediate returns – they have to be made for the long game. Also, the costs of borrowing money need to be taken into account – borrowing is both expensive and can be risky if repayment dates are not met. Our industry needs motorcycling enthusiasts who can see the long-term benefits of manufacturing motorcycles.

Fixed costs – These are the costs that need to be covered each month, irrespective of the number of motorcycles that are made. These will include building costs – rent, business rates, fixed energy costs, maintenance, and fixed staff costs.

Variable costs – These are the costs to make each bicycle, mainly materials and direct labor. As the output of motorcycles or components increases, the variable costs will increase too. Of course, if increased production is because of increased sales, the revenue will increase too if the pricing is constant. The management challenge in running any manufacturing business is

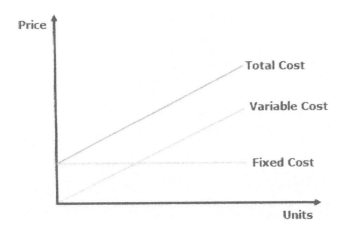

Figure 12.2 Break-even graph.

the balancing of costs and sales to ensure passing the **break-even point,** that is where the revenue exceeds both the fixed costs and the variable costs and the business is making a surplus – in other words, a profit.

Sometimes companies have rapid growth, expand, and so increase costs; then a slump in sales occurs and the company is out of business because it cannot meet its costs. This happens in the motorcycle industry sometimes, but many old companies have kept steady by controlling growth and maintaining steady production. A well-documented example from the automotive industry is the Morgan Motor Company; they manufacture a range of sports cars. Morgan has been making cars for over 110 years. Their production rate remains fairly constant, but their waiting list time changes. As I write this, the Morgan waiting list is six months; it has been as long as ten years. It shows that if you make a good product, customers will wait for delivery.

PRODUCTION PLANNING

Aircraft producer Boeing has a rolling 3-year production schedule for their works in Seattle. Customers are advised of this when placing orders, but some changes to specifications can be made before production starts. Some motorcycle manufacturers also have similar production schedules.

Mass production – This is about producing large numbers of identical components. Mass production originally started in about 1861 at the start of the American Civil War. The war saw a large number of guns being made to satisfy both Union and Confederate armies. The gun makers developed the mass production method. These methods were then used in the production of other items, which, of course, included motorcycles.

The key points of mass production are:

- Large quantity of identical parts is used
- Each person specializes in doing only one part of the manufacturing process
- Components are designed for ease of assembly
- A flow line, or assembly line, is used to move the product between the different stages of manufacture

This tends to lead to lots of identical motorcycles being made. Motorcycles produced in this way tend only to be available in limited range of engine sizes. They also usually have a very limited paint color range. When you consider the numbers of motorcycles being made, it is easy to see why the ranges are limited and the same components are used throughout the range.

Handmade craft production – In contrast to mass production, there is the making of individual motorcycles for individual customers. Every single detail can be tailored for the rider who is buying a motorcycle. The price of these handmade machines is about ten times that of the mass-produced ones. Companies such as Harris Performance and Yoshimura make specialist parts and convert standard motorcycles to race machines. Manufacturers such as Bimota modify parts and make frames and bodywork to make original machines.

This handcraft production, which usually involves CNC (Computer numerical control) machined parts and hand assembly, is now big business. Lots of new motorcycle buyers want their machine to be original in some way, so custom, one-off machines are highly desirable.

Offshore production – To cut costs and therefore increase profits, many motorcycle manufacturers have their components manufactured outside their home country. Production tends to be centered in the Far East, that is China and Taiwan. It is worth noting that China and Taiwan have close connections politically and economically, although there are many differences culturally.

ASSEMBLY METHODS: MODERN RETAILING

Retailing of motorcycles used to be about the motorcycle shop owner knowing his or her customers, stocking items known to sell, and ordering from catalogues nonstocked items. Your local shop would offer great service based on specialist knowledge. Then in some areas, current retailing is more like "stack them high and sell them cheap," as depicted in the novel *The Long Firm* by Jake Arnott. The concentration is on sales volume, profit margin, and delivery times. There are many forms of retailing, including:

- High street retail motorcycle shops
- Café-style shops – an increasing growth of destination venues
- eBay/Gumtree or similar based shops
- Department stores – especially for clothing and accessories
- Retail/business park–based shops
- Shops at events and shows
- Online and magazine advertised mail order

All these businesses operate on what is called the commercial, or commerce, model. It is about buying as cheaply as possible and selling for as high a price as possible. In commerce, both buying and selling can be problematic. Generally, in the motorcycle industry the problem is sourcing the right product at the right price to be able to sell at a profit after sales costs. Cost of sales includes:

- Advertising
- Transport
- Post and packaging
- Stocking/warehousing
- Depreciation
- Losses and deterioration
- Cost of premises, rent, and rates
- Staffing costs

Fashion – Like any other consumer product, the motorcycle is subject to changes in fashion. Currently the big fashion is the modern take on retro. Currently naked, traditional-style machines account for over 40% of all sales. To go with these naked machines the popular jackets are Barbour and Belstaff, which are based on 1930s designs.

THE WAY WE LIVE

How we live is going to influence future motorcycle sales. With smaller houses and flats and people with less time and fewer practical skills, there is likely to be more emphasis on retailers to provide more services alongside their sales.

We also have safety trousers, which look like jeans and boots that can be worn anywhere. With scooter riders, we have modern versions of parka coats and 1960s-style helmets that meet modern safety requirements.

220 Motorcycle Engineering

Figure 12.3 Sunday morning ride-outs to destinations are very popular; some dealers have installed coffee bars and fast food outlets to attract riders to their premises.

OTHER MOTORCYCLE INDUSTRY SERVICES

Classic restorations – The growing interest in classic motorcycles stems from riders wanting machines that they couldn't afford, or that otherwise weren't accessible to them, when they were younger. The word classic covers just about any era and any machine. The emphasis used to be on veteran – built before 1904 – and vintage – built between 1904 and 1931. Then there was the term post-vintage, meaning 1931 up to World War II.

We now call 1990s and later motorcycles classics – and even modern-day classic.

There is a very large industry related to the supply of parts and services for classics, usually specializing in one particular make and model. I know of one company that has a 2-year waiting list for the restoration of a single make of 1960s racing engines.

Race services – This can take two main forms: providing technical support for a particular team or providing what is called neutral services. Big race teams such as those with BSB (British Super Bike) or WSB (World Super Bike) status have their own full-time dedicated service team. The service team will service and repair the motorcycles throughout the whole year, ensuring that they are perfect for the rider from the early season practice sessions to the final end of season appearance events. This includes cleaning the motorcycles and transporting them between locations.

Lesser rated professional/semi-professional teams contract in service only for the events which they consider major to their race calendar. So, the service crew tends to work elsewhere during the rest of the year.

There are also neutral service crews that provide services to anybody, or team, at an event. This may be part of a publicity campaign to promote their products to riders and the public.

Specialist parts and services – With motorcycles and their components becoming more complex there is an increasing need for specialist parts and services. The expense and complexity of items such as fuel injectors, anti-lock braking systems (ABS), and carbon fiber wheels require a large investment in both stock and special tools, and, of course, the specially equipped van to transport them in.

Mobile services – A number of small businesses are setting up to provide mobile services, typically visiting set locations at set times. This may be to a country town market, a busy shopping mall, or a parking lot in a small village. With the increasing costs of business rent and rates, the mobile business is a very attractive low-cost business concept.

Holidays – A growth area is the motorcycling holiday market; this can take several forms, such as:

- Arrive and ride – motorcycles provided, meet at airport, accommodation and meals provided.
- Own motorcycle – transfer service and accommodation.
- Fully guided tours with **breakdown wagon** – a following van for repairs or tired riders.
- Use of sat-nav which is preset to act as a guide.

Touring – Like the holiday concept, it takes different forms; the main one is that a van, acting also as a breakdown wagon, carries the luggage between the different night stops. This is motorcycle touring made easy.

Transportation – Many motorcyclists like to tour different parts of the country and do not want to ride to the area, nor wish to use a car. The transport of bicycles on trains and airplanes can be problematic. So, using specialist transport, taking care of the motorcycle and luggage, provides a good answer. This business is also growing in the motorcycle market, particularly for city-based riders who want to ride off-road in areas such as Scotland.

Bicycle concierge – This concept is about motorcycling for the very busy business person, having the motorcycle and kit ready when he or she gets off the airplane. This service is very popular for supercar owners – a Ferrari waiting in the valet parking area, but you can have your Brough Superior waiting with riding kit and helmet.

Make and sell direct/custom bicycles – This area of the motorcycle business is growing, with an increasing number of older people with high disposable income in most countries. This is not about named brands of bicycles, but about functionality – meaning adventure motorcycles and super touring models.

Hobby market/paraphernalia/fan-zone/collectables – Providing those bit and bobs that sit on our shelves and hang on our walls. The author collects items related to racing and classic motorcycling. He often finds himself bidding against friends who collect the same range of items.

Motorcycle hire – A growing part of the holiday and leisure industry across the world. Flat seaside towns and country areas with unpaved roads offer particularly good opportunities because riding is easy and safe. The favored make is Harley Davidson, as they are fairly bulletproof. The Beach Boys sing about renting a Honda in one of their 1960s surfing songs. Charges are usually by the hour.

Accessories – Accessory sales depend on having an attractive item, at a good price, and marketing it correctly. It's all about the unique selling point (USP).

TRADE ASSOCIATIONS

There are three well-known trade associations that offer services to the cycle industry in the UK:

- Motorcycle Industry Association (MCIA)
- National Motorcycle Dealers Association (NMDA)
- Federation of Small Businesses – not specifically about motorcycles, but gives a wide range of services and support to any small business

Most countries have similar associations. These are set up to give advice and guidance to businesses and to lobby politically for the benefit of such businesses.

Support from these associations variously includes but is not limited to:

- Technical training
- Reduced-cost finance and banking
- Advertising materials
- E-commerce solutions
- Tax advice
- Legal advice
- Professional business promotion

As membership is not expensive, it is normal to be a member of more than one association.

Chapter 13
Reinforced composite materials

INTRODUCTION

Composite materials came into use in the automotive and boat building industries in the 1950s to fulfil the needs of small postwar car manufacturers, as metal was rationed for 12 years after the end of World War II and small companies sought alternatives. The automotive market was developing at the same time as the small boat market; both developed along the same lines using a glass fiber and a resin lay-up procedure called glass-reinforced plastics (**GRPs**), which is still used today by many kit car manufacturers and makers of boats and some motorcycle parts. The use of **composites – a product made up of more than one material that are bonded together to provide special properties** – became more specialized as **carbon-based technical materials** became available. Carbon fiber, as it is called, was originally invented and patented by the Royal Aircraft Establishment at Farnborough, UK. Bicycles started to be made from it in the 1990s as the materials became more readily available and methods of joining tubes and laying up became known. The usage of carbon fiber for motorcycles has until recently been limited to small parts, such as huggers. Now manufacturers are starting to experiment with it as a frame material. Given the use of molds and autoclave equipment, the manufacture of frames in carbon fiber should be cheaper and require less skill than the construction of frames from metal.

Depending on the materials used in composite construction, the following properties of composites may influence their choice:

- Components can be produced on a one-off basis with minimum tooling
- Compound curvature can be produced with constant material thickness
- Extreme lightness for a given strength
- Resistant to corrosion
- Different finishes are available
- There is a style cache in the use of carbon fiber

BASIC PRINCIPLES OF REINFORCED COMPOSITE MATERIALS

The basic principle involved in reinforced plastic production is the combination of polyester resin and reinforcing fibers to form a solid structure. GRPs are essentially a family of structural materials that utilize a wide range of thermoplastic and thermosetting resins. The incorporation of glass fibers in the resins changes them from relatively low-strength, brittle materials into strong and resilient structural materials. In many ways glass fiber–reinforced plastic can be compared to concrete, with the glass fibers performing the same function as the steel reinforcement and the resin matrix acting as the concrete. Glass fibers have high strength and high modulus, and the resin has low strength and low modulus. Despite this the resin has the important task of transferring the stress from fiber to fiber, so enabling the glass fiber to develop its full strength.

Polyester resins are supplied as viscous liquids, which solidify when the actuating agents, in the form of a catalyst and accelerator, are added. The proportions of this mixture, together with the existing workshop conditions, dictate whether it is cured at room temperature or at higher temperatures and also the length of time needed for curing. In common practice pre-accelerated resins are used, requiring only the addition of a catalyst to affect the cure at room temperature. Glass reinforcements are supplied in a number of forms, including chopped strand mats, needled mats, bidirectional materials such as woven rovings and glass fabrics, and rovings that are used for chopping into random lengths or as high-strength directional reinforcement. Other materials needed are the releasing agent, filler, and pigment concentrates for the coloring of glass fiber–reinforced plastic.

Among the methods of production, the most used method is that of contact molding, or the wet laying-up technique, as it is sometimes called. The mold itself can be made of any material that will remain rigid during the application of the resin and glass fiber, which will not be attacked by the chemicals involved and which will also allow easy removal after the resin has set hard. Those in common use are wood, plaster, sheet metal, and glass fiber itself, or a combination of these materials. The quality of the surface of the completed molding will depend entirely upon the surface finish of the mold from which it is made. When the mold is ready the releasing agent is applied, followed by a thin coat of resin to form a gel coat. To this a fine surfacing tissue of fiber glass is often applied. Further resin is applied, usually by brush, and carefully cut-out pieces of mat or woven cloth are laid in position. The use of split washer rollers removes the air and compresses the glass fibers into the resin. Layers of resin and glass fibers are added until the required thickness is achieved. Curing takes place at room temperature, but heat can be applied to speed up the curing time. Once the catalyst has caused the resin to set hard, the molding can be taken from the mold.

MANUFACTURE OF REINFORCED COMPOSITE MATERIALS

When glass is drawn into fine filaments, its strength greatly increases over that of bulk glass. Glass fiber is one of the strongest of all materials. The ultimate tensile strength (UTS) of a single glass filament (diameter 9 to 15 micrometers) is about 3,447,000 kN/m^2. It is made from readily available raw materials and is noncombustible and chemically resistant. Glass fiber is therefore the ideal reinforcing material for plastics. In Great Britain, the type of glass that is principally used for glass fiber manufacture is E glass, which contains less than 1% alkali borosilicate glass. E glass is essential for electrical applications, and it is desirable to use this material where good weathering and water resistance properties are required. Therefore it is greatly used in the manufacture of composite body shells.

Basically the glass is manufactured from sand or silica, and the process by which it is made proceeds through the following stages:

1. Initially the raw materials, including sand, china clay, and limestone, are mixed together as powders in the desired proportions.
2. The "glass powder," or frit as it is termed, is then fed into a continuous melt furnace or tank.
3. The molten glass flows out of the furnace through a forehearth to a series of fiberizing units, usually referred to as bushings, each containing several hundreds of fine holes. As the glass flows out of the bushings under gravity, it is attenuated at high speed.

After fiberizing the filaments are coated with a chemical treatment usually referred to as a forming size. The filaments are then drawn together to form a strand, which is wound on a removable sleeve on a high-speed winding head. The basic packages are usually referred to as cakes and form the basic glass fiber which, after drying, is processed into the various reinforcement products. Most reinforcement materials are manufactured from continuous filaments ranging in fiber diameter from 5 to 13 micrometers. The fibers are made into strands by the use of size. In the case of strands that are subsequently twisted into weaving yarns, the size lubricates the filaments as well as acting as an adhesive. These textile sizes are generally removed by heat or solvents and replaced by a chemical finish before being used with polyester resins. For strands that are not processed into yarns, it is usual to apply sizes that are compatible with molding resins.

TYPES OF REINFORCING MATERIAL

Woven fabrics

Glass fiber fabrics are available in a wide range of weaves and weights. Lightweight fabrics produce laminates with higher tensile strength and modulus than heavy fabrics of a similar weave. The type of weave will also influence the strength (due, in part, to the amount of crimp in the fabric), and satin weave fabrics, which have little crimp, usually give stronger laminates than plain weaves, which have a higher crimp. Satin weaves also drape more easily and are quicker to impregnate. Besides fabrics made from twisted yarns, it is now the practice to use woven fabrics manufactured from rovings. These fabrics are cheaper to produce and can be much heavier in terms of weight.

Chopped strand mat

Chopped strand glass mat (**CSM**) is the most widely used form of reinforcement. It is suitable for molding the most complex forms. The strength of laminates made from chopped strand mat is less than that made with woven fabrics, since the glass content that can be achieved is considerably lower. The laminates have similar strengths in all directions because the fibers are random in orientation. CSMs consist of randomly distributed strands of glass about 50 mm long, which are bonded together with a variety of adhesives. The type of binder or adhesive will produce differing molding characteristics and will tend to make one mat more suitable than another for specific applications.

Needle mat

This is mechanically bound together, and the need for an adhesive binder is eliminated. This mat has a high resin pick-up due to its bulk and cannot be used satisfactorily in molding methods where no pressure is applied. It is used for press molding and various low-pressure techniques such as pressure injection, vacuum and pressure bag.

Rovings

These are formed by grouping untwisted strands together and winding them on a "cheese." They are used for chopping applications to replace mats either in contact molding (spray-up) or translucent sheet manufacture of press molding (preform). Special grades of roving are available for each of these different chopping applications. Rovings are also used for weaving, for filament winding, and for pultrusion processes. Special forms are available to suit these processes.

Chopped strands

These consist of rovings prechopped into strands of 6-mm, 13-mm, 25-mm, or 50-mm lengths. This material is used for dough molding compounds and in casting resins to prevent cracking.

Staple fibers

These are occasionally used to improve the finish of moldings. Two types are normally available: a compact form for contact molding and a soft bulky form for press molding. These materials are frequently used to reinforce gel coats. The weathering properties of translucent sheeting are considerably improved by the use of surfacing tissue.

Application of these materials

CSM is most commonly used for the average molding. It is available in several different thicknesses and specified by weight: 300, 450, and 600 g/m^2. The 450 g/m^2 is the most frequently used and is often supplemented with the 300 g/m^2. The 600 g/m^2 density is rather too bulky for many purposes and may not drape as easily, although all forms become very pliable when wetted with the resin. The woven glass fiber cloths are generally of two kinds, made from continuous filaments or from staple fibers. Obviously, most fabricators use the woven variety of glass fiber for those structures that are going to be the most highly stressed. For example, a molded glass fiber seat pan and squab unit in an HPV (Human Powered Vehicle) would be made with woven material as reinforcement, but a detachable hard top for a trailer body would more likely be made with CSM as a basis. However, it is quite customary to combine cloth and mat not only to obtain adequate thickness, but because if the sandwich of resin, mat, and cloth is arranged so that the mat is nearest to the surface of the final product, the appearance will be better.

The top layer of resin is comparatively thin, and the weave of cloth can show up underneath it, especially if some areas have to be buffed subsequently. Chopped fibers do not show up so prominently, but some fabricators compromise by using the thinnest possible cloth (surfacing tissue as it is known) nearest the surface, on top of the CSM. When molding, these orders are, of course, reversed, with the tissue going on to the gel coat on the inside of the mold, followed by the mat and resin lay-up.

It is important to note that if glass cloths or woven mat is used, it is possible to lay up the materials so that the reinforcement is in the direction of the greatest stresses, thus giving extra strength to the entire fabrication. In plain weave cloths, each warp and weft thread passes over one yarn and under the next. In twill weaves, the weft yarns pass over one warp and under more than one warp yarn; in 2 × 1 twill, the weft yarns pass over one warp yarn and under two warp yarns. Satin weaves may be of multishaft types, when each warp and weft yarn goes under one and over several yarns. Unidirectional cloth is one in which the strength is higher in one direction than the other, and balanced cloth is a type with the warp and weft strength about equal. Although relatively expensive, the woven forms have many excellent qualities, including high-dimensional stability, high tensile and impact strength, good heat, weather and chemical resistance, moisture absorption, resistance to fire, and good thermoelectrical properties. A number of different weaves and weights is available, and thickness may range from 0.05 mm to 9.14 mm, with weights from 30 g/m^2 to 1 kg/m^2, although the grades mostly used in the automotive field probably have weights of about 60 g/m^2 and will be of plain, twill, or satin weave.

Carbon fiber

This is another reinforcing material. Carbon fibers possess a very high modulus of elasticity and have been used successfully in conjunction with epoxy resin to produce low-density composites possessing high strength.

RESINS USED IN REINFORCED COMPOSITE MATERIALS

The first manmade plastics were produced in this country in 1862 by Alexander Parkes and were the forerunner of celluloid. Since then, a large variety of plastics has been developed commercially, particularly in the last 25 years. They extend over a wide range of properties. Phenol formaldehyde is a hard thermoset material; polystyrene is a hard, brittle, thermoplastic; polythene and plasticized polyvinyl chloride (PVC) are soft, tough thermoplastic materials; and so on. Plastics also exist in various physical forms. They can be bulk solid materials, rigid or flexible foams, or in the form of sheet or film. All plastics have one important common property. They are composed of macromolecules, which are large chain-like molecules consisting of many simple repeating units. The chemist calls these molecular chains polymers. Not all polymers are used for making plastic moldings. Manmade polymers are called synthetic resins until they have been molded in some way, when they are called plastics.

Most synthetic resins are made from oil. The resin is an essential component of glass fiber–reinforced plastic. The most widely used is unsaturated polyester resin, which can be cured to a solid state either by catalyst and heat or by catalyst and accelerators at room temperature. The ability of polyester resin to cure at room temperature into a hard material is one of the main reasons for the growth of the reinforced plastics industry. It was this which led to the development of room-temperature contact molding methods which permit production of extremely large integral units.

> **Tech Note**
>
> Scientists are working on making carbon fiber materials biodegradable.

Polyester resins are formulated by the reaction of organic acids and alcohols, which produces a class of material called esters. When the acids are polybasic and the alcohols are polyhydric, they can react to form a very complex ester which is generally known as polyester. These are usually called alkyds and have long been important in surface coating formulations because of their toughness, chemical resistance, and endurance. If the acid or alcohol used contains an unsaturated carbon bond, the polyester formed can react further with other unsaturated materials such as styrene or diallyl phthalate. The result of this reaction is to interconnect the different polyester units to form the three-dimensional cross-linked structure that is characteristic of thermosetting resins. The available polyesters are solutions of these alkyds in the cross-linking monomers. The curing of the resin is the reaction of the monomer and the alkyd to form the cross-linked structure. An unsaturated polyester resin is one that is capable of being cured from a liquid to a solid state when subjected to the right conditions. It is usually referred to as polyester.

CATALYSTS AND ACCELERATORS

In order to mold or laminate a polyester resin, the resin must be cured. This is the name given to the overall process of gelation and hardening, which is achieved either by the use of a catalyst and heating or at normal room temperature by using a catalyst and an accelerator. Catalysts for polyester resins are usually organic peroxides. Pure catalysts are chemically unstable and liable to decompose with explosive violence. They are supplied therefore as a paste or liquid dispersion in a plasticizer or as a powder in an inert filler. Many chemical compounds act as accelerators, making it possible for the resin-containing catalyst to be cured without the use of heat. Some accelerators have only limited or specific uses, such as quaternary ammonium compounds, vanadium, tin, or zirconium salts. By far the most important of all accelerators are those based on a cobalt soap or those based on a tertiary amine. It is essential to choose the correct type of catalyst and accelerator, as well as to use the correct amount, if the optimum properties of the final cured resin or laminate are to be obtained.

Pre-accelerated resins

Many resins are supplied with an in-built accelerator system controlled to give the most suitable gelling and hardening characteristics for the fabricator. Pre-accelerated resins need only the addition of a catalyst to start the curing reaction at room temperature. Resins of this type are ideal for production runs under controlled workshop conditions.

The cure of a polyester resin will begin as soon as a suitable catalyst is added. The speed of the reactions will depend on the resin and the activity of the catalyst. Without the addition of an accelerator, heat, or ultraviolet radiation, the resin will take a considerable time to cure. In order to speed up this reaction at room temperature, it is usual to add an accelerator. The quantity of accelerator added will control the time of gelation and the rate of hardening.

There are three distinct phases in the curing reaction:

Gel time. This is the period from the addition of the accelerator to the setting of the resin to a soft gel.

Hardening time. This is the time from the setting of the resin to the point when the resin is hard enough to allow the molding or laminate to be withdrawn from the mold.

Maturing time. This may be hours, several days, or even weeks depending on the resin and curing system, and is the time taken for the molding or laminate to acquire its full hardness and chemical resistance. The maturing process can be accelerated by post-curing. When the material is not fully matured, it is referred to as being in its **green state**, or simply as **green**; this term is taken from the color of the wood of a freshly chopped down tree.

Fillers and pigments

Fillers are used in polyester resins to impart particular properties. They will give opacity to castings and laminates; produce dense gel coats; and impart specific mechanical, electrical, and fire resisting properties. A particular property may often be improved by the selection of a suitable filler. Powdered mineral fillers usually increase compressive strength; fibrous fillers improve tensile and impact strength. Molding properties can also be modified by the use of fillers; for example, shrinkage of the molding during cure can be considerably reduced. There is no doubt, also, that the wet lay-up process on vertical surfaces would be virtually impossible if thixotropic fillers were not available.

Polyester resins can be colored to any shade by the addition of selected pigments and pigment pastes, the main requirement being to ensure thorough dispersion of coloring matter throughout the resin to avoid patchy moldings.

Both pigments and fillers can increase the cure time of the resin by dilution effect, and the adjusted catalyst and promoter are added to compensate.

Releasing agents

Releasing agents used in the normal molding processes may be either water-soluble film-forming compounds or some type of wax compound. The choice of releasing agent depends on the size and complexity of the molding and on the surface finish of the mold. Small moldings of simple shape, taken from a suitable GRP mold, should require only a film of polyvinyl alcohol (PVAL) to be applied as a solution by cloth, sponge, or spray. Some moldings are likely to stick if only PVAL is used. PVAL is available as a solution in water or solvent, or as a concentrate which has to be diluted, and it may be in either colored or colorless form.

Suitable wax emulsions are also available as a releasing agent. They are supplied as surface finishing pastes, liquid wax, or wax polishes. The recommended method of application can vary depending upon the material to be finished. Hand apply with a pad of damp, good-quality mutton cloth or equivalent in straight, even strokes. Buff lightly to a shine with a clean, dry, good-quality mutton cloth. Machine at 1800 rpm using a G-mop foam finishing head. Soak this head in clean water before use and keep damp during compounding. Apply the wax to the surface. After compounding, remove residue and buff lightly to a shine with another clean, dry, good-quality mutton cloth.

Wax polishes should be applied in small quantities since they contain a high percentage of wax solids. Application with a pad of clean, soft cloth should be limited to an area of approximately 1 square meter. Polishing should be carried out immediately before the wax is allowed to dry. This can be done either by hand or by machine with the aid of a wool mop polishing bonnet.

Frekote is a semi-permanent, multirelease, gloss finish, nonwax polymeric mold release system specially designed for high-gloss polyester moldings. It will give a semi-permanent release interface when correctly applied to molds from ambient temperature up to 135 °C. This multirelease interface chemically bonds to the mold's surface and forms on it a micro-thin layer of a chemically resistant coating. It does not build up on the mold and will give a high-gloss finish to all polyester resins, cultured marble, and onyx. It can be used on molds made from polyester, epoxy, metal, or composite molds. Care should be taken to avoid contact with the skin, and the wearing of suitable clothing, especially gloves, is highly recommended. These products must be used in a well-ventilated area.

Adhesives used with GRP

Since polyester resin is highly adhesive, it is the logical choice for bonding most materials to GRP surfaces.

Suitable alternatives include the Sika technique, which is a heavy-duty, polyurethane-based joining compound. It cures to a flexible rubber which bonds firmly to wood, metal, glass, and GRP. It is ideal for such jobs as bonding glass to GRP or bonding GRP and metal, as is often required on HPVs with GRP bodywork. It is not affected by vibration and is totally waterproof. The Araldite range includes a number of industrial adhesives which are highly recommended for use with GRP. Most high-strength impact adhesives (superglues) can be used on GRP laminates.

Most other adhesives will be incapable of bonding strongly to GRP and should not be used when maximum adhesion is essential.

Core materials

Core materials, usually polyurethane, are used in sandwich construction, that is basically a laminate consisting of a foam sheet between two or more glass fiber layers. This gives the laminate considerable added rigidity without greatly increasing weight. Foam materials are available that can be bent and folded to follow curved surfaces such as motorcycle parts. Foam sheet can be glued or stapled together then laminated over to produce a strong box structure, without requiring a mold. Typical formers and core materials are paper rope, polyurethane rigid foam sheet, scoreboard contoured foam sheet, Termanto PVC rigid foam sheet, Term PVC contoured foam sheet, and Termino PVC contoured foam sheet.

Formers

A former is anything that provides shape or form to a GRP laminate. They are often used as a basis for stiffening ribs or box sections. A popular material for formers is a paper rope, made of paper wound on flexible wire cord. This is laid on the GRP surface and is laminated over to produce reinforcing ribs, which give added stiffness with little extra weight. The former itself provides none of the extra stiffness; this results entirely from the box section of the laminate rib. Wood, metal, or plastic tubing and folded cardboard can all be used successfully as formers. Another popular material is polyurethane foam sheet, which can be cut and shaped to any required form.

Composite theory

In its most basic form, a composite material is one which is composed of two elements working together to produce material properties that are different from the properties of those elements on their own. In practice, most composites consist of a bulk material called the matrix and a reinforcement material of some kind which increases the strength and stiffness of the matrix.

Polymer matrix composite (PMC) is a type of composite used in modern vehicle bodywork. This type of composite is also known as fiber-reinforced polymers (or plastics) (FRPs). The matrix is a polymer-based resin, and the reinforcement material is a fibrous material such as glass, carbon, or aramid. Frequently, a combination of reinforcement materials is used.

The reinforcement materials have high tensile strength, but are easily chaffed and will break if folded. The polymer matrix holds the fibers in place so that they are in their strongest position and protects them from damage.

The properties of the composite are thus determined by:

- The properties of the fiber
- The properties of the resin
- The ratio of fiber to resin in the composite – **fiber volume fraction (FVF)**
- The geometry and orientation of the fibers in the composite

Resin

The choice of resins depends on a number of characteristics, namely:

- Adhesive properties – In relation to the type of fibers being used and if metal inserts are to be used such as for panel fitting.
- Mechanical properties – Particularly tensile strength and stiffness.
- Micro-cracking resistance – Stress and age hardening causes the material to crack; the micro-cracks reduce the material strength and eventually lead to failure.
- Fatigue resistance – Composites tend to give better fatigue resistance than most metals.
- Degradation from water ingress – All laminates permit very low quantities of water to pass through in a vapor form. If the laminate is wet for a long period, the water solution inside the laminate will draw in more water through the osmosis process.
- Curing properties – The curing process alters the properties of the material. Generally oven curing at between 80 °C and 180 °C will increase the tensile strength by up to 30%.
- Cost – The different materials cost different prices.

The main types of resins are polyesters, vinylesters, epoxies, phenolics, cyanate esters, silicones, polyurethanes, bismaleides (**BMIs**), and polyamides. The first three are the ones mainly used for manufacturing work, as they are reasonably priced. Cyanates, BMI, and polyamides cost about ten times the price of the others.

Reinforcing fibers

The mechanical properties of the composite material are usually dominated by the contribution of the reinforcing fibers. The four main factors that govern this contribution are:

1. The basic mechanical properties of the fiber
2. The surface interaction of the fiber and the resin – called the interface
3. The amount of fiber in the composite – **FVF**
4. The orientation of the fibers

The three main reinforcing fibers used in HPVs are glass, carbon, and aramid. In addition, the following are used for nonbody purposes: polyester, polyethylene, quartz, boron, ceramic, and natural fibers such as jute and sisal.

Aramid fiber is a manmade organic polymer, an aromatic polyamide, produced by spinning fiber from a liquid chemical blend. The bright golden yellow fibers have high strength and low density, giving a high specific strength. Aramid has good impact resistance. Aramid is better known by its Dupont trade name: Kevlar.

Carbon fiber is produced by the controlled oxidation, carbonization, and graphitization of carbon-rich organic materials – referred to as precursors – which are in fiber form. The most common precursor is polyacrylonitrile (PAN); pitch and cellulose are also used.

PRE-IMPREGNATED MATERIAL (PRE-PREG)

Woven material is available pre-impregnated with resin. It is referred to as pre-preg. This means that the material has exactly the right amount of resin applied to it. The resin is fully coating the material so that there are no dry spots that could lead to component failure. Pre-preg is therefore quicker to use, and the resin density is accurate.

Pre-preg has a limited shelf life, which is compounded by the fact that it must be stored at -18 °C. A deep freeze cabinet is therefore needed for storage. The pre-preg cannot be unrolled or cut when it is in the frozen state, so it must be removed from the freezer and brought up to normal room temperature. It is only possible to freeze and defrost the pre-preg a limited number of times, so the material must be managed carefully. The usual way to do this is by means of a control card. The dates and times of defrosting are recorded, as is the amount of material taken off the roll. That way the life of the roll and the amount of material left can be seen without removing the roll from the freezer.

Curing

The resin, whether it is by wet lay-up or pre-preg, needs time and heat to dry it out and make it hard. When the hardener is added to the resin, it will generate heat chemically. Be careful – this heat can cause fire and other damage. However, at normal temperature, 20 °C, it will take about 5 days for the resin to become fully hard. During this period the component should not be moved nor should any stress be applied. To speed up the hardening process and to add extra strength to the component, it is normal to use an oven. The oven may be a simple box with an heating element or an autoclave, which is a cylindrical shaped oven that can be pressurized or evacuated inside. The normal procedure is to place the newly made component in the oven or autoclave then rack up the temperature gently, over a period of about 30 minutes. Maintain the temperature typically at 150 °C for about 5 hours, then gradually lower the temperature, again over about a 30-minute period. The best way to do this is with a computer control system.

Core materials

Engineering theory tells us that in most cases the stiffness of a panel is proportional to the cube of its thickness. That is, the farther apart that we can keep the outer fibers, the stiffer the panel will be. Putting a low-density core between two layers of composite material will add stiffness with minimum weight and at reasonable cost.

Foam

A variety of materials are used – one of the most common is foam. Foam can be made from a variety of synthetic polymers Densities of foam can vary between 30 and 300 kg/m^3 and thicknesses available are from 5 to 50 mm.

Honeycomb

Honeycombs are made from a variety of materials, including extruded thermoplastic – ABS, polycarbonate, polypropylene, and polyethylene – bonded paper, aluminum alloy, and for fire-resistant parts, Nomex. Nomex is a paper-like material based on Kevlar fibers.

Heat

A point to be noted is that most carbon fiber materials are affected by heat. Thermal expansion can lead to micro-cracking. A carbon fiber panel that is painted black will absorb a lot of heat if left in the sun for a long period. This can cause the panel to expand, which could lead to micro-cracks in the panel and cracks in the paint work. This will then allow in moisture, which will cause further deterioration of the panel.

Chapter 14
Data

Just about every individual and organization collect data of some sort or another. At a simple level, it may be checking the instruction sheet to see what the tire pressures should be or collecting a power readout from the dynamometer.

> **Tech Note**
>
> Data, noun plural: facts and statistics collected together for reference or analysis. The singular is datum. Data include statistics and information.

Motorcycle riders collect data for such things as mileage covered or the best time for a lap of the Nürburgring. However much more data are available for riders and company owners, for example:

- Power output and torque curves
- Tightening torques
- Tire pressure
- Compression pressures
- Sales statistics

THE NORM

The norm is something which is usual, typical, or standard.

The secret in the use of data is actually knowing what they show and what you can do with that information. In other words, what is the norm and what are the extremes, what makes them these cases, and how can we utilize this information. We'll start with an explanation of the norm. This concept came about through population studies – scientists visiting other countries many years ago and comparing the height and other attributes of the inhabitants of these countries. They would then say the norm for the height of people from country A is x so that they could compare them in country B, which is y. So, they developed the concept of the normal distribution curve. That is to say, not all people in country A are the same height, but the bulk of the population will vary within a few centimeters either side of x. In addition, there will be both very tall people and very short ones – those outside the norm. If you go into any High Street clothes shop, you'll find that they usually only stock a limited range of sizes – the ones which fit those within the norm. Often this range is simply small, medium, and large in their own ratings and varies with manufactures.

UNDERSTANDING AND USING DATA

To use data, it is a good idea to consider Bloom's Taxonomy. It states the six levels of the cognitive domain. Most people work in the lower three levels: **knowledge, comprehension,** and **application** in Bloom's terminology. In other words, understanding the description of the data, being able to describe and discuss the actual content of the data, and being able to apply the actual data in a real situation.

The three higher levels are **analysis, synthesis,** and **evaluation.** Analysis means breaking the data down, understanding both the content and the structure. In terms of your race results, it is considering all the factors: motorcycle used, gearing, clothing, course, weather, distance, time of day, other riders, and what you had for breakfast, among a myriad of other items. Synthesis is picking out the bits which you think were most important and remixing them to form a new structure. Evaluation is making a judgement on whether the new structure worked – then making more changes to improve it again. It is a continuous process.

Recording data

Recording data is very important: how you record it will control the way in which you can use it. Businesses such as takeaway shops use data to control their opening hours to match their customer needs. For instance, rarely will you find a chip shop open before 12 p.m., but London kebab shops often stay open till 4 a.m.

Don't waste time on recording data which you will not use, and remember that if they data are not recent, they won't be relevant. Mood, fashions, and tastes change quickly, as does technology and many other factors.

Tally Chart for Colours at a show

Colour	Tally	Frequency
Red	⊥⊥⊥⊢	5
Orange	I I I	3
Yellow	I I	2
Green	I I I I	4
Blue	I I I I	4
Inigo	I I	2
Violet	I	1

Figure 14.1 Tally chart, count in blocks of five.

Tally chart

A simple way of collecting data is the use of a tally chart. The first column has the options listed. The second color is the tally; this is made in pencil line strikes up to four, then the fifth is across. The third column is the tally totaled.

Figure 14.2 Manual clicker – good for counting footfall numbers at events.

If it is the number of attendees at an event, a simple manual tally counter is ideal. Push the lever for each head counted. The total number will be displayed.

Spreadsheet

Spreadsheets are a great way of recording data, and, of course, you can carry out detailed analysis, synthesis, and evaluation using *what if* scenarios on them. That is inserting projected values in the columns and seeing what happens to the results. This is of particular value when you are concerned with data changes of very small margins, say less than 1%. For example, when changing your gearing by one tooth and retaining the same engine speed, how much will this shave off your 0 to 60 mph time? What will the extra profit be when increasing the price of something by a few pence? To give you an indication of the value of spreadsheets, F1 teams use them to work out how to shave half a second off lap times with their £120M annual budget.

Presenting data

Spreadsheets are good for the recording and the analysis of data, but to get your message across, to both yourself and to others, a visual form is useful. Print them out and pin them up on the wall to help motivate yourself and your staff or colleagues.

BAR CHARTS AND STACKING BAR CHARTS

The bar chart and stacking bar chart are good to show growth and changes in emphasis. The best approach is probably to make them in color.

PIE CHARTS

The pie chart is a really good way of showing percentage data, as it gives a clear indication of percentages. A circle is made up of 360 degrees, so 10% is 36 degrees. To find, for example, 30% as an angle for the pie chart:

$$30/100 \times 360° = 108°$$

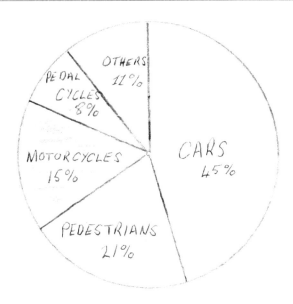

Figure 14.3 Pie chart.

NORMAL DISTRIBUTION

Calculations for normal distribution are simple to do. The following gives you two examples; ungrouped and grouped data use different methods. It is useful to be able to calculate the mean (average), the variance from the mean, and the standard deviation.

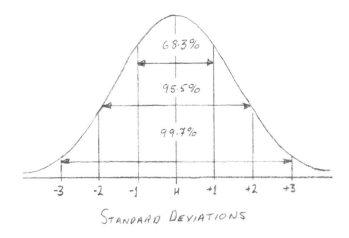

Figure 14.4 Normal distribution chart showing standard deviations.

Ungrouped data

The following is a table of voltages measured across the lighting circuit of a low-voltage system in a motorcycle workshop to check volt drop.

Sample	Voltages	Differences from Mean	Differences Squared
1	119	1.46	2.13
2	120	0.46	0.211
3	120	0.46	0.211
4	120	0.46	0.211
5	121	0.54	0.291
6	119	1.46	2.13
7	122	1.54	2.37
8	122	1.54	2.37
9	123	2.54	6.45
10	123	2.54	6.45
11	119	1.46	2.13
12	118	2.46	6.05
13	120	0.46	0.211
Total: 13	Total: 1566		Total: 31.215

Mean:

$$\bar{x} = \frac{Total\ of\ voltages}{Number\ in\ sample} = \frac{1566}{13} = 120.46$$

Variance:

$$S = \frac{Sum\ of\ differences\ squared}{Number\ in\ sample} = \frac{31.215}{13} = 2.401$$

Standard deviation σ:

$$\sqrt{S} = sigma\ \sigma = \sqrt{2.401} = 1.549$$

Grouped data

You are carrying out a quality check on suppliers. The table gives the sizes in mm of samples of 100-mm gear shafts.

Line	Range/group mm	Midpoint x	Frequency f	fx	$\bar{x}$	$(x - \bar{x})$	$f(x - \bar{x})^2$
1	89–91	90	17	1530	99.07	-9.07	1398.5
2	91–93	92	18	1656	99.07	-7.07	899.7
3	93–95	94	19	1786	99.07	-5.07	488.4
4	95–97	96	20	1920	99.07	-3.07	188.5
5	97–99	98	30	2940	99.07	-0.07	34.3
6	99–101	100	50	5000	99.07	0.93	4326
7	101–103	102	35	3570	99.07	2.93	300
8	103–105	104	22	2288	99.07	4.93	535
9	105–107	106	18	1908	99.07	6.93	864
10	107–109	108	10	1080	99.07	8.93	89
			Total: 239	Total: 23678			Total: 9123.4

Mean:

$$\bar{x} = \frac{\Sigma(fx)}{\Sigma f} = \frac{23678}{239} = 99.07$$

Variance:

$$S = \frac{f(x - \bar{x})2}{\Sigma f} = \frac{9123.4}{239} = 38.17$$

Standard deviation σ

$$\text{Sigma } \sigma = \sqrt{S} = \sqrt{38.17} = 6.178$$

PROFIT AND LOSS ACCOUNT (PLA)

This is essential data to see if the company is actually making a profit or a loss. With the growth of community interest companies (CICs,) generally not-for-profit organizations, these need to be crystal clear to show all incomes and outgoings. This is essential to assure the HMRC and others where the money is going.

ANY MOTORCYCLE COMPANY

Profit and Loss Account
  Income £
  Motorcycle sales* 25,000
  Accessory sales* 4000
  Service work 3000
  Repair work 10,000
  *Less purchase costs
  Total 42,000
  Expenses £
  Rent of premises 5000
  Business rates 1000
  Electricity 250
  Water 300
  Gas 300
  Transport 1500
  Technology 250
  Total 8600
  Trading Profit £33,400

Balance sheet

The balance sheet sets out the company's position with the relationship between what it owns and what it owes at any one point in time. To keep abreast of the trading, it is normal to produce a balance sheet for each quarter of the year and compare it to previous quarters. The motorcycle industry has variation with the seasons, so each quarter results may be very different, but the comparison of any quarter with those of the previous year's same quarter should give a good indication of the trading situation.

Acid-test ratio (**ATR**) – This is the absolute test of a company's viability. It is calculated by dividing the company's total current assets (TCA) by the total current liabilities (TCL).

$$ATR = TCA / TCL$$

For the 2 years shown in the balance sheet, these are:

$$2019\ ATR = 204{,}000 / 50{,}000 = 4.08$$

$$2020\ ATR = 222{,}500 / 50{,}000 = 4.45$$

ANY CYCLE COMPANY

Balance Sheet
2019 2020

	£	£
Current Assets		
Cash in hand	1500	500
Cash in bank	10,500	13,000
Stock at valuation	120,000	130,000
30-day assets	20,000	15,000
Inventory assets	10,000	12,000
Subtotal	16,200	170,500
Long-term assets		
Tools and equipment	10,000	7000
Goodwill	2000	5000
Premises	30,000	40,000
Subtotal	42,000	52,000
Total current assets	204,000	225,000
Current liabilities		
Bills payable	30,000	32,000
Loans payable	10,000	8000
Tax payable	10,000	10,000
Total current liabilities	50,000	50,000

It can be seen that the ATR has improved as the company has matured. If the ATR is greater than unity (1), the company is viable; if less then unity, there is likely to be a need to borrow money.

Some examples of data you might collect

Distances covered, fuel used, lap times, service intervals, business data, helmet and clothing usage for safety. There is a range of data collection hardware available, but that is for a future book.

Chapter 15
Science terminology

The history of motorcycles shows that they are about the application of science and technology. The companies with experience in precision engineering from the manufacture of guns were the ones to succeed first, with BSA and Royal Enfield being two of the most well-known names. Then along came Honda, who refined the science in the manufacturing process, thus working to even finer limits and getting double the power output from their engines compared to the traditional manufacturers and adding features nobody else had without increasing the weight. This chapter will help you understand the terminology surrounding the science related to motorcycles.

SI SYSTEM AND COMMON UNITS

SI stands for Système International, a system of measurement units that was developed following World War II. There are several different systems of measurement in use throughout the world, but for the examinations with UK-based examining bodies, SI only is used. It is worth noting that in countries such as Germany and Japan they use SI but with amendments and modifications. The Germans use Deutsch (Germany) Industrial Norm (DIN). The Japanese use Japanese Industrial Standards (JIS). The Americans use American National Standards Institute (ANSI), as well as SI. We'll also discuss some of the others that are used in America too – for instance, the imperial system – so called after the British Empire of the Victorian era was used in the UK up until the 1970, so it is found on most classic motorcycles. With variations, this is still used in the United States alongside SI.

The **imperial system of measurement** for length uses inches, feet, yards, and miles. For mass, it uses ounces, pounds, and tons.

Table 15.1 SI units

Quantity	Quantity Symbol	Unit	Unit Symbol
Length	l	meter	m
Mass	m	kilogram	kg
Time	t	second	s
Electric Current	I	ampere	A
Temperature	T	kelvin	K

The following table is an **approximate** guide.

Please note that the Glossary and List of Abbreviations section of this book gives more information about units and related terminology.

The engineering convention is to measure tube lengths center to center. That is the center of one end lug to the center of the other. However, in bicycle seat tubes, lots of makers measure them from the center of the bottom bracket to the top of the seat tube. This may in fact include an extension above the top tube, especially with inclined top tubes.

DECIMALS AND ZEROS

It's very easy to make mistakes with decimal calculations and the use of zeros. As you will have seen from Table 15.2, the standard or basic units are often too big or too small in value. So, a series of multiples and sub-multiples is used to make calculations easier. For instance, kilo – meaning thousand – is added to meter to give kilometer. In other words, 1000 meters. Going in the other direction, we use milli – meaning one-thousandth – when

Table 15.2 Imperial/SI approximations

Quantity	Imperial	SI
Length	1 inch (in)	25 mm
Length	1 foot (ft)	300 mm
Length	1 yard (yd)	900 mm
Length	39 inches	1 meter
Length	1 mile	1.6 kilometers
Mass	1 ounce	25 grams
Mass	1 pound	454 grams
Mass	2.2 pounds	1 kilogram (kg)
Mass	1 ton	1000 Kilogram

Table 15.3 Multiples and Submultiples

Prefix	Symbol	Power	Number
giga	G	10^9	1,000,000,000
mega	M	10^6	1,000,000
kilo	k	10^3	1000
hecto	h	10^2	100
deca	da	10^1	10
deci	d	10^{-1}	0.1
centi	c	10^{-2}	0.01
milli	m	10^{-3}	0.001
micro	µ	10^{-6}	0.000001

we are talking about the very low voltages in vehicle electronics, for example, millivolts.

ACCURATE MEASURING

Accurate measuring – called metrology in engineering terminology – is used extensively in motorcycle manufacture. Older machines were made with much larger tolerances than current ones. Tight accurate tolerances give:

1. Improved motorcycle quality and appearance – enabling the machine to be sold for the highest possible price.
2. Reduction in wind noise – giving a more enjoyable ride.
3. Improved aerodynamics – thus improving performance.
4. Enabling more economical use of materials – thus reducing manufacturing costs

Five pieces of equipment that are used in vehicle manufacture are starting to find their way into motorcycle manufacture. These are currently scientific instruments and require skilled usage, but easy-to-use versions are now becoming available. Let's briefly discuss each one:

Coordinate Measuring Machine (CMM) – When we are making a brake bracket or steerer tube to fit a frame, the first important measurements will be the mounting points. All other measurements will be taken from these points which we call coordinates – like the points on a map. A CMM measures distances from coordinates, even on the most irregular shaped object such as a scooter frame. It does this to an accuracy of one micron (1µm, a millionth of a meter). This is necessary, as often the first prototypes are made by hand from the contours of clay models – called a buck. Therefore, the exact measurements may not be known.

Granite Block – This is what its name says: a gigantic block of granite on which a motorcycle can be placed. This block of granite will have a mass greater than that of the motorcycle and will be supported on many hydraulic columns so that it can be kept perfectly level. Even the most level road surface will have a natural curvature – the curvature of the Earth. This granite block is made dead square. It enables dead accurate measurements to be taken of the motorcycle – measurements which can be used during the construction process. Frame construction is usually undertaken on jigs – sets of metal rails that support the frame members, with attachments to position the fittings and hangers. The measurements from the granite block are used to inform the measurements for the production jig. For quality sampling, complete motorcycles can be measured on the granite blocks.

Lasers – Can be used for positioning component mountings during construction. It is important to have accurate frame measurements using the CMM and position measurements from the granite block. It is possible to transfer this information to the frame jig to ensure that the components are correctly positioned before fixing takes place – either welding or bonding. This is done by setting up one or more lasers on the outer part of the jig that will shine a light, or several spots will converge together, when the part is accurately positioned. The part can then be bonded or welded into place most accurately.

Scanners – Measuring an irregularly shaped object, such as a fuel tank, can be done without starting to mark the object; then the measurements may be transferable to metal to make another one. This has two main uses. In manufacturing the scanner can be used to take the profile from the design mockup or first hand-made metal part and convert it into a CADCAM file for the manufacture of a press die. The other is to scan the profile of vintage motorcycle parts to aid in manufacturing new ones.

Wind tunnel – Increasingly used to measure drag.

$$\text{The Drag Force} = \tfrac{1}{2} \text{ air density} \times \text{velocity squared} \times \text{frontal area} \times \text{drag co-efficient}$$

This might seem like a lot of variables, but it isn't really. Most wind tunnels will give some form of read out of force, or you can attach streamers to actually see what is happening. You can then alter the wind speed by turning the fan motor up. Altering the rider position or fairing shape will change the frontal area, and changing the shape or material will alter the drag coefficient.

CAPACITY AND VOLUME

Liquids in the UK and Europe are sold in liters – this can be in parts of a liter or multiples of liters; for example, a half-liter or maybe 5 liters. In the United States it may be sold in pints or gallons.

A liter is defined as the volume of 1000 cubic centimeters – 1000 cc. Water has a mass of 1 kg per liter at a temperature of 4 °C. This is often referred to as density. Oil and other lubricants are lighter than water and have a density of about 0.9 kg per liter.

Sometimes the term Relative Density (RD) or Specific Gravity (SG) is used. Both words mean the same thing. That is the density of the oil is compared to the density of water. So, the oil – irrespective of its volume – if 1 liter weighs 0.9 kg will have an RD of 0.9.

Still used by some companies, and more so in the United States, are pints and gallons. Beer is also frequently sold in pints. Be aware that English pints and gallons are different from American ones.

English pints are made up of 20 fluid ounces – a little obscure – that means the volume of 20 ounces of water at 17 °C. An English pint of water weighs a pound and a quarter. An English gallon is 8 English pints weighing 10 pounds. It is defined as 4.54 liters.

American gallons are defined as 231 cubic inches, that is 3.78 liters. An American gallon of water weighs 8.34 lb. So, it is considerably smaller. An American pint is 16 fluid ounces.

The volume changes with temperature – the volume of both oil and beer are legally measured in the UK at 16 °C.

TEMPERATURE AND HEAT

These two scientific terms are often misused, so let's get them cleared up so that you know what you are talking about when welding or brazing.

> *Temperature* – This is the hotness or coldness of an object. There are three temperature scales in use:

Celsius (C) – Also known as Centigrade because it has 100 degrees in it. It is related to the freezing and boiling point of water. Water freezes at 0 °C) and boils at 100 °C.

Fahrenheit (F) – Water freezes at 32 °F and boils at 212 °F.

Kelvin (k) – This unit just uses the letter k and no degree symbol. It is the absolute temperature scale. 0 °C equals 273 k. 100 °C equals 373 k. Absolute zero, which is the lowest temperature achievable, is 0 k, which equals –273 °C.

Heat – This is the amount of energy used to raise the temperature. Heat is a form of energy. It takes 4200 joules of heat energy to raise the temperature of 1 kg of water 1 °C. Water is said to have a specific heat of 4200 J/kg C.

When you are using a gas torch to warm up something, for instance, a piece of metal that you wish to bend, you will note that it takes time. Different metals take different times, and larger pieces take longer than smaller ones of the same metals. The longer time means that it is using more gas – this means more heat. As an example, typically 1 kg of propane will give 50 MJ.

FORCE AND PRESSURE

These two terms are also often confused or misused. Let's clarify them – it'll come in useful when you are pushing and bending or straightening something.

Force – We often use this in calculations but don't necessarily understand it. The unit of force is the newton (N), named after Sir Isaac Newton who first discovered it. He noticed that if anything was dropped, it would go to the ground. This is due to the force of gravity (G). The farther an item drops, the faster it goes – this is called acceleration due to gravity. The rate of acceleration on Earth is typically 9.81 m/s/s. For terms of simple calculations, we often use 10 m/s/s as the equivalent of G:

$$\text{Force(N)} = \text{Mass(kg)} \times \text{Acceleration(G)}$$

Tech Note

Mass is another name for weight when we are doing calculations with earth-bound objects.

So, imagine an average person with mass 65 kg stood on the roof of a car:

$$\text{Force of person's feet on car roof} = 65\,\text{kg} \times 10\,(\text{value of G})$$
$$= 650\,\text{N}$$

That is a static force. If we have a dynamic force – say swinging a hammer – then to get an approximate value, we call it 2 sigma; in other words, we double the total force. So, for a large sledge hammer of 5.5 kg (12 lb) with a moderate swing, we get 2(5.5 × 10) = 110 N.

Pressure – This is force divided by the cross-sectional area:

$$\text{Pressure}\,(N/m^2) = \text{Force}\,(N) / \text{Cross-sectional area}\,(m^2)$$

To avoid confusion with other units, the term pascal (Pa) is used for 1 N/m² pressure.

The pressure of 1 pascal is very low – imagine an apple on your desktop – that's about 1 Pa. So, we tend to use the term bar – this is equivalent to normal barometric pressure. One bar equals 101.3 KPa. The air compressor that provides air pressure for your power tools generates about 10 bar in pressure. Motorcycle tires are typically inflated to between 2.5 and 3 bar.

AMPS, VOLTS, OHMS, WATTS, AND KIRCHHOFF

Electricity is easy to understand if you get the basic terms clear. Although you can't see electricity, it behaves in a similar way to water. Providing that the plumbing in your workshop is connected to the mains supply, when you turn the tap on, water will flow out. Water comes out under pressure – a pressure from the main supply. The consumer standard for water pressure in the UK is enough pressure to fill a 4.5-liter bucket in 30 seconds. Typically, about 2 bar.

When we talk about electricity, the switch replaces the tap, the voltage (V) replaces the pressure, and the amps (I) replaces the full bucket. So, the voltage needs to be high enough to force enough amps through to light up your lamp, for example. Items in a circuit provide a resistance (R); like the tap, they can slow or stop the flow of electricity.

SAFETY NOTE

ELECTRICITY CAN KILL YOU!
Please be aware that any voltage or amperage of electricity can kill you. It can also give you a nonlethal shock, which can make you flinch or jump and cause personal injury by hitting a rotating part or a hot part.

Ohm's law – This is the relationship between amps (I) and volts (V), which will give a value for the resistance measured in ohms (R).

$$I = V / R$$

Watt (W) – This is a measure of power. If you look at light bulbs, they will usually have their power rating on them; the same applies to electric motors and heating elements. It is normal to state the power and the voltage on these electrical items. With LED lights, the equivalent wattage is often given in two ways. For instance, some mains LED lights are rated as 9 W = 100 W. That is, they consume 9 W but give the equivalent light of a 100 W tungsten bulb – the older type of light bulb. They still work off a 230V mains power supply.

$$Watts(W) = Volts(V) \times Amps(I)$$

Tech Note

I = current in amps.

A is often used in a colloquial or less formal way and also means amps.

If you have the wattage and the voltage, you can work out the current consumption in amps.

$$I = W / V$$

Knowing the current consumption is useful when you are finding faults or wiring up a new component. When finding faults, you can use an induction ammeter to see the actual current flowing. When fitting a component, knowing current consumption allows you to choose the correct cable size.

Kirchhoff – At any junction in an electrical circuit, the current flowing into the junction will equal the current flowing out. This gives you more information when testing a circuit – finding where the current is flowing to.

SAFETY NOTE

Always isolate the circuit concerned and disconnect the battery where appropriate.

FRICTION

Skidding – This happens when the friction between the tires and the road is not sufficient to keep the motorcycle on a course. Friction is the ratio between the force acting downwards on the tire (weight [W]) and the force (F) needed to slide the tire over the road – not rolling the tire, but making it skid. On a normal road, with good tires, this ratio, µ (the Greek letter mu), expressed as a decimal fraction, is about 0.8. If the road is covered in ice or wet leaves, the ratio can be as low as 0.01 – in other words, it can be pushed along with its brakes on.

µ = Force / Weight

SOME COMMON LAWS OF MECHANICAL ENGINEERING

Newton's Laws are about force and acceleration. They are numbered as follows:
   First Law – A body – motorcycle, piece of metal, etc. – will either stay where it is or continue moving uniformly unless another force is applied to it.
   Second Law – The force on a body is equal to its mass multiplied by its rate of acceleration. Usually expressed in the form Force (F) = Mass (m) × Acceleration (a).
   Third Law – When a force is applied to a second body, the second body will be exerting a force backwards. In other words, for each action there is an equal and opposite reaction.
   Hooke's Law – This is about how metal reacts to force. Metal stretches by an amount (X) proportionally to the force (F) applied to it until it gets to its elastic limit, when it breaks. The amount of stretch depends on the type of metal; the metal type is given a constant (k).

$F = k \cdot X$

IMPACT AND MOMENTUM

Momentum – When a motorcycle is traveling along a road, it processes momentum – Newton's First Law. That momentum (p) is the product of its mass (m) and velocity (v) – that is, speed combined with direction.

$p = mv$

The heavier the vehicle and the faster the speed, the greater the momentum. This is why we have speed limits.
   Kinetic energy – We know that the moving vehicle possess momentum, and again applying Newton's Laws – this time the Second Law – we have to apply a force to stop.

Chapter 16

Transmission

This chapter is about the getting the power and torque transmitted from the engine to the back wheel.

CLUTCH

The purpose of the clutch is to transmit the torque, or turning force, from the engine to the transmission. It is designed so that the drive can be engaged and disengaged smoothly and easily. By disengaging the drive, the clutch allows the gears to be changed smoothly, and it provides a temporary neutral position. This allows the transmission gears to be engaged or disengaged while the engine is running.

The clutch assembly is contained in the housing at one end of the gearbox.

Key points

- The main components of the clutch are the pressure plate, the spinner plate(s), and the thrust bearing.
- Clutches may have several friction plates.
- Clutch dust is a potential health hazard.
- Clutch adjustment is important to ensure that it does not slip.

Transmission of torque

The transmission of torque from the engine to the gearbox depends on the strength of the springs in the pressure plate, the diameter of the spinner plate, the number of friction surfaces, and the coefficient of friction of the clutch materials. The stronger the springs and the larger the diameter of the spinner plate, the greater the torque that can be transmitted. The number of friction plates in motorcycle clutches varies – typically six, which means 12 friction surfaces.

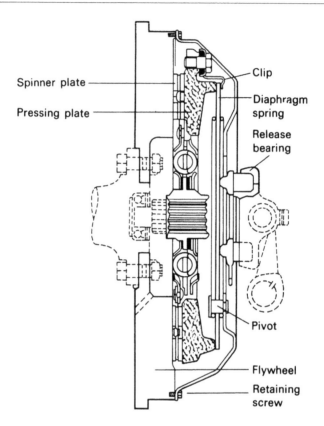

Diaphragm-spring clutch engaged

Figure 16.1 Diaphragm spring clutch engaged – motorcycles may have up to six spinner plates and pressing plates.

The types of springs are coil springs used on older bikes and diaphragm springs used on most modern ones. The problem with coil springs is that if one or two go weak, you get clutch judder. This does not happen with diaphragm springs.

Diaphragm-spring clutch

The diaphragm spring is shaped like a saucer or a deep dinner plate, with a series of radially cut grooves.

In the engaged position the diaphragm spring is shaped like a saucer. The force of the outer rim forces the pressure plate against the spinner plate. The pressure plate cover is bolted to the flywheel, and the pressure plate is attached to the cover with flexible metal straps. When the flywheel rotates, the cover rotates – the cover turns the pressure plate by means of the metal

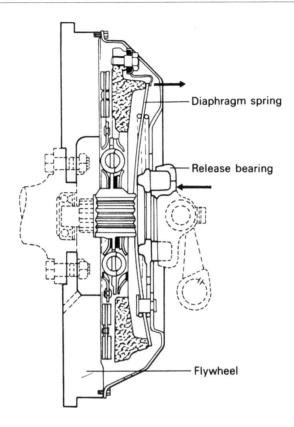

Diaphragm-spring clutch disengaged

Figure 16.2 Diaphragm spring clutch disengaged.

straps. Most engines rotate in a clockwise direction when viewed from the right, as the straps must pull, not shove; thus, the clutch must be assembled so that the strap pulls in a counterclockwise direction when seen from the other side.

The diaphragm spring pivots on the cover using rivets with the shoulders as fulcrum points. A fulcrum is another name for a pivot – something to swing on. It is against these rivets that the spring forces itself to transmit force to the pressure plate and the flywheel to transmit the drive.

To disengage the clutch, the thrust race presses in the middle of the diaphragm spring. This causes the spring to pivot on the shoulders of the rivets so that it lifts its outer rim. This is similar to the action of a jam jar lid or a CD in its case. The lifting of the outer rim pulls the pressure plate away from the spinner plate so that the spinner plate can rotate freely. The clutch is now disengaged so that the drive is not transmitted to the gearbox.

Spinner plate

The spinner plate consists of a **steel hub** which fits on to the splines of the gearbox input shaft. Attached to the hub is a disc that carries the friction material. The friction material is riveted to the disc.

Cable clutch

The cable clutch is used on many motorcycles. The inner twisted wire Bowden cable moves inside a steel outer guide cable (sometimes called a sheaf). The lever pulls the inner cable, which in turn moves the clutch cross-shaft. The outer cable acts to provide a guide and controls the length for adjustment purposes. This system is similar to a bicycle brake or gear cable; it is adjusted in a similar way, by using a screwed nipple to change the length, thus taking out the slack. The flexibility of the cable means that vibrations from the engine are unlikely to be transmitted to the clutch pedal.

Hydraulic clutch

A more sophisticated system is the hydraulic clutch. This system uses hydraulic fluid, like that used in hydraulic brakes, to transmit the movement from the clutch lever to the cross-shaft in the clutch housing. This system has a mechanical advantage: the difference between the force applied by the driver's hand on the lever and the actual force that the clutch thrust race applies to the pressure plate diaphragm is built into the hydraulic system.

The clutch lever moves a push rod, which in turn pushes a hydraulic piston into the clutch master cylinder. The hydraulic fluid above the piston is forced along the connecting tube. In turn, the fluid forces the clutch cylinder piston against a short operating rod, which transmits the force to the clutch cross-shaft. These systems are very smooth in terms of operation, being used on most bikes. It is essential to ensure that the clutch master cylinder reservoir is kept topped up to the correct level with the correct type and grade of hydraulic fluid. Most bikes use the same fluid for both the brakes and the clutch.

NOMENCLATURE

Mechanical advantage is the leverage gain, or torque multiplication, given by a mechanical mechanism. For instance, you might need to apply a force of 50 N to the clutch lever and move the pedal 40 mm to disengage the clutch. The linkages in the clutch mechanism may have increased this force to at the thrust race to 500 N and reduced the distance traveled to 4 mm. The mechanical advantage would be 10 to 1.

Clutch adjustment

It is essential that there is enough **free-play** in the clutch mechanism for the thrust bearing to be clear of the pressure plate. This is needed to ensure that the clutch does not slip. It is also important that there is not too much free play; otherwise, the clutch may not be able to be disengaged completely.

The normal amount is 0.5 or 1 mm of free play at the thrust race or cross-shaft lever.

Clutch faults

The main faults likely to occur on a clutch are slipping and grabbing. Slipping is when the clutch is not transmitting the drive; it might be brought about by:

- The spring being fatigued or tired
- Oil on the friction lining
- The friction lining being worn down to the rivets
- Lack of free play

Grabbing is when the clutch cannot be engaged smoothly. That is, the clutch suddenly grabs and takes the drive up with a thud. Grabbing may be caused by:

- Spring damage or uneven wear
- Wear in the mechanism or friction lining
- A broken spinner plate hub

Transmission

> **Key points**
>
> - The transmission includes the gearbox, the final drive gears, the propeller shaft, and/or the drive shafts.
> - The gearbox allows the engine speed to be varied to suit the road conditions.
> - Straight cut, helical, double helical, and epicyclic gears are used.

Component layout

On older motorcycles with pre-unit construction, the gearbox is mounted behind the engine; power is transmitted from the engine to the gearbox by a primary chain. With unit construction, there is a gear drive from the engine to the gearbox.

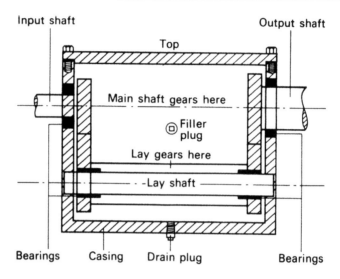

Basic layout of gearbox

Figure 16.3 Layout of basic gearbox.

GEARBOX

Function

The job of the gearbox is to allow the motorcycle to accelerate and climb hills easily. This is done by using a selection of gear trains that enable changes to be made in the ratio of engine speed to wheel speed.

Principles of gearing

The reason for needing a gearbox is that the engine only develops usable power over a limited range of speeds – called the power band. The speed at which power is developed depends on the type of engine. For instance, the maximum power of the four-cylinder Kawasaki ZX6R is 129 HP at 13,500 RPM, while the Yamaha Bulldog, which is nearly twice the capacity with two cylinders, produces 65 HP at 5500 RPM.

The gearbox acts like a lever, enabling a small engine to move a very heavy object. This is like how a tire lever enables the tire fitter to apply great force to the tire bead.

> ### Tech Note
> The gearbox has three main functions; it provides:
>
> 1. Low gears for acceleration, moving heavy loads, and climbing steep gradients
> 2. High gears to enable high-speed cruising
> 3. A neutral gear, so that the engine can be running while the car is stationary

Gear ratio

The gear ratio of any two meshing gears is found by the formula:

Gear ratio = Number of teeth on driven gear/Number of teeth on driver gear

This is usually written = Driven / Driver

Where two gears mesh together, the gear ratio is:

Gear ratio = B/A
= 50 / 25
= 2 / 1

This is written 2:1 (read 2 to 1).

This means that for each two turns of A, B will rotate one turn; hence two (turns) to one (turn). That is gear B will rotate at half the speed of gear A. In other words, B will rotate at half the number of revolutions per minute compared with gear A.

If equal-size pulleys and ropes were attached to the shafts to which gears A and B are fixed, it would be possible to use the 10-kg (22-lb) weight to balance the 20-kg (44-lb) weight. This is because the turning effort, or torque, is increased proportionally to the gear ratio. Although the speed is halved, the turning effort is doubled. This effect of the gear ratio is used when climbing steep hills or pulling heavy loads, such as a trailer.

Gearbox ratios

> ### Tech Note
> Gear ratios are a bit confusing at first; it is a good idea to look at a sectioned gearbox to understand what is happening in the metal, as you might say. If you cannot see a sectioned gearbox, try to find an old gearbox that you can take apart. Also, most workshop manuals have lots of pictures of gearboxes, which might help you understand how the gears run together.

Compound gear ratio = Driven/Driver * Driven/Driver

In our example:

= B / A * D / C

If the number of teeth on each wheel is:

A = 10

B = 20

C = 15

D = 30

The formula would become:

Gear ratio = 20/15 * 30/15
= 600/150
= 4/1
= 4 : 1

Final drive gear ratio

The final drive ratio is the ratio of the speed of the gearbox output to that of the road wheel.

The final gear ratio = Number of teeth on rear sprocket/Number of teeth on gearbox sprocket

For example, with a 50-toothed rear sprocket and a 10-toothed gearbox sprocket:

Final drive ratio = 50/10
= 5 : 1

Overall gear ratio

The overall gear ratio is the ratio of the speed of the engine to the speed of the road wheels. This is found by multiplying the gearbox ratio by the final drive ratio:

Overall gear ratio (OGR) = Gearbox ratio * Final drive ratio

If the gearbox ratio is 2.5:1 and the final drive ratio is 3:1, then:

OGR = Gearbox ratio * Final drive ratio
= 2.5 * 3
= 7.5 : 1

Gear teeth

There are three main types of gear teeth in use gearboxes: spur gear, helical gear, and double helical gear. Each kind of gear can be identified by the shape of its teeth.

Spur gear – This has straight teeth, like a cowboy's spur. This type of gear is also called straight cut. Straight cut gears can only carry a limited

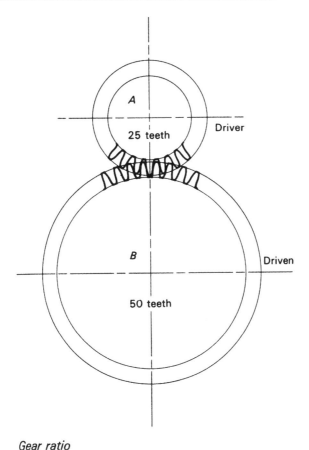

Gear ratio

Figure 16.4 Gear ratio.

load, and they are noisy in operation. You can hear straight cut gears whilst and occasionally rattle.

Helical gear – So called because if the shape of these teeth were projected, as around a long tube, the shape formed would be a helix. Another example of a helix is the screw thread on a bolt. Because the tooth is longer than the gear is wide, it is stronger than the equivalent straight cut gear.

Double helical gears – Are made like two rows of opposing helical gears. This is machined from one piece of metal so that the side thrust on one half of the gear balances the side thrust on the other half. Hence, there is no tendency of the gear to move sideways on the main shaft.

268 Motorcycle Engineering

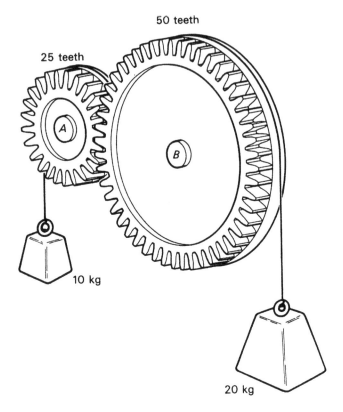

Gear ratio: turning effort

Figure 16.5 Turning effort: 10 kg balances 20 kg.

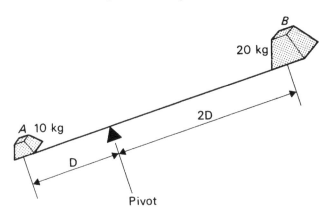

Gear ratio showing lever principle

Figure 16.6 Lever principle.

Spur gear

Figure 16.7 Spur gear.

Helical gear

Figure 16.8 Helical gear.

Double helical gear

Figure 16.9 Double helical gear.

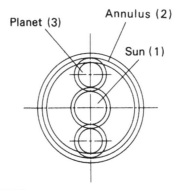

Epicyclic gear

Figure 16.10 Epicyclic gear.

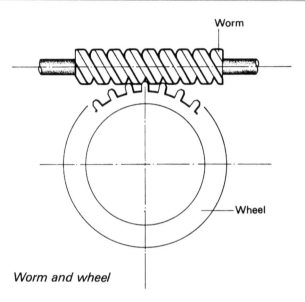

Figure 16.11 Worm and wheel gear.

Figure 16.12 Gear pedal on lightweight motorcycle.

Transmission 271

Figure 16.13 Separate gear box casing – nonunit construction.

Figure 16.14 Cover of primary drive chain on nonunit construction.

272 Motorcycle Engineering

Figure 16.15 Chain adjuster on lightweight motorcycle.

Figure 16.16 Chain adjuster on sports tourer motorcycle.

Transmission 273

Figure 16.17 Notice crash bobbin on end of axle and bobbin for pit stand.

Figure 16.18 Nice carbon fiber plate to protect rider's boots from chain.

Figure 16.19 Toothed belt drive on a Zero Electric Motorcycle – similar to that used on Harley Davidson.

Chapter 17

Tuning and customizing

MODIFICATIONS

> **CAUTION**
>
> *If you carry out any modifications to any vehicle, whatever happens is your responsibility.*

Health and safety and the environment

Like any other task, you must be aware of health, safety, and environmental issues.

Legal implications

It is important to be aware of the legal implications of any modifications that you make to a vehicle. It is not illegal to sell many parts, but when they are fitted to a road-going vehicle, this could be illegal. Also, on competition vehicles, the parts must comply with the requirements of the racing regulations relating to that particular type of racing or specific class. As a technician, under corporate law, which is vicarious by its nature, if a vehicle to which you have fitted a part is involved in an incident and the part that you fitted might have been a cause of the damage, then you may be held wholly or partially liable for the damage caused. That damage could be the death of an innocent person, in which case you could be charged with manslaughter.

So, think carefully about the modifications you are making. Do not just do it because a customer asked you to do it. Customers are often unaware of the laws, and indeed racing regulations. It is your duty to advise and guide them; you have a duty of care to your customers.

276   Motorcycle Engineering

Figure 17.1 Indian – almost the size of a car.

Figure 17.2 Chrome plated V-twin with foot boards.

Figure 17.3 Car-like dashboard.

Figure 17.4 Gloriously chromed front end.

278 Motorcycle Engineering

Table 17.1 Common Mistakes in Modifications

Area	Modification	Mistake	Comment
Tires	Racing tires	Slicks with no tread	Illegal on road
Tires	Asymmetric tires	Wrong direction of rotation	Unsafe in wet conditions
Wheels	Wide wheels	Extending beyond guards	Illegal on road or track
Lighting	Fitting spot or fog lamps	Incorrect height or position	Illegal on road
Lighting	Any lamps moved by changes to body work	Moved or restricted vision	Check that they comply with position on vehicle and for angle of vision
Exterior fittings	Bolts or screws on exterior of body work	Must have a minimum of 2 mm radius and no sharp edges	Common reason for kit cars failing SVA
Exhaust	Change system or parts	Noise limits	Especially appropriate to bikes
Ignition	Change parts	May alter engine operation	Could affect emissions
Fuel system	Change parts	May change fuel emissions	Check within limits
Brakes	Changing pads	Could alter braking characteristics	Lots of riders are surprised to find the extra effort needed with competition brake pads
Engine	Fit a bigger/more powerful one	Need to upgrade the brakes and suspension too	I admit to doing this and wondered why it took a long time to stop (very dangerous!)
Suspension	Fit lowered springs	Need to fit shock absorbers to suit	Suspension will bottom

Some of those points should make you smile – but they are all potentially dangerous and most seriously to be avoided; think the modification through, think safe.

Points to be noted when preparing a race engine:

- Manufacturers sometime use different blocks for the same engine in different applications; variations include materials and manufacturing processes.

- Sometimes one block is used for a variety of different engine capacities – the capacity is changed by using a crank shaft with a different stroke.
- There are several different types of block construction.
- The same material can have different properties based on the manufacturing process; for instance, aluminum can be sand-cast or die-cast. The former is more resilient for race engines, although the later might be lighter in terms of mass.

Factors affecting the exhaust efficiency and sound are:

- Manifold design and subsequent system.
- Material used for exhaust – some vintage motorcycles use copper, and this produces a delightful resonance.

Figure 17.5 Orange and black.

- Use, or otherwise, of a catalytic converter – motorcycles operate best at about 900 °C.
- Use of a turbo-charger.

Reasons for modifications

The reasons why people modify or enhance their vehicles are many and varied, but usually they can be classified as one of the following:

- Making it faster
- Making it more powerful
- Making it lighter
- Making it handle better

Figure 17.6 Slipped checkered pattern, CNC forward-mounted foot pegs, and drilled brake pedal.

- Making it stop more quickly
- Making it more comfortable to ride – especially under specific conditions, such as extra lights for night driving
- Making it look good and more attractive
- Making it comply with specific racing regulations

A good motorcycle technician will know all the right tweaks to enable the bike to perform better than the other competitors.

Cylinder head

If you modify the cylinder head, you will increase the pressure and heat generated, so you need to consider the following:

- Hoses – high pressure
- Hose clips – high clamping pressure
- Coolant – inhibitor for use with aluminum
- Use of wetting agent
- Radiator flow capacity
- Coolant (water) pump – flow rate
- Coolant pump drive belt and velocity ratio
- Cylinder head gasket – check material, thickness, and use of rings
- Cylinder head studs or bolts – need high-tensile-strength ones

Valves

The purpose of the valve is to open and close to control gas flow. Its opening and closing speed is limited by its mass – remember Newton's Second Law, $F = Ma$. So, the valve needs to be as light in weight as possible; for this reason titanium is frequently used. The higher the engine output, the higher the amount of heat generated. So the valves need to be able to dissipate heat. For this reason, sodium-filled valves are sometimes used, as the latent heat of the sodium filling serves to reduce the valve temperature.

Valve shape is also of consideration in terms of valve seats and head shape to improve the flow of the gases both into and out of the cylinder head.

Valve spring

The valve springs close the valves. The camshaft opens the valves, so the faster the engine revs, the faster they will open. Remember that if we can make the engine rev faster, it will probably be able to develop more power – within certain restrictions. So, we are now dependent on the springs closing the valves; if we are increasing the engine speed, we will need to fit stronger valve springs. Otherwise, the springs will not close the valves fast enough and the engine will suffer from valve bounce – indeed, the valves may touch the piston crown. To modify the valve spring closing rate, there are usually three options: stronger single springs, double springs, or if double springs are already fitted, triple springs can be fitted.

Fitting double or triple springs may only be possible if the valve retaining caps are changed to match the new springs, and the spring seat on the cylinder head may need machining to accommodate the new springs.

Camshaft

The camshaft controls the valve opening in terms of both lift and period:

- Valve lift – The distance the valve is lifted from its seat. Like opening a door, the greater the lift, the more gas that can be put through at any one point of time. The valve lift may be the same as the lift at the cam, or there may be a ratio so that the valve lift is greater than the cam lift when using a rocker mechanism. Direct-acting cams do not have this option.
- Period (also called duration) – The number of degrees that the valve is open, measured between the opening and the closing points. The number of degrees will correspond to a period of time for any given engine speed. Again – like a door – the longer it is open, the more petrol and air that can get through it. If the valve period is increased from, say, 100 degrees to 110 degrees, it will be open 10% longer, so 10% more gas can pass through it. This is often simply referred to as valve timing or just timing.

There are three main options:

1. Change the camshaft for one with different opening periods and/or lift.
2. Change the rocker mechanism ratio.
3. Change the cylinder head and camshaft for one in a different location – for instance, fitting a double overhead cam (DOHC) cylinder head in place of an overhead valve (OHV) arrangement.

Also to be considered is the material from which the camshaft is made. Many popular vehicles use cast iron because it is easily made, cheap, and sufficiently hard for normal usage. For competition use, it is normal to use a high grade of steel with sufficiently high carbon content so that it can be induction hardened.

Figure 17.7 Black with ape-hanger bars.

Cylinder block

Cylinder blocks are usually the subject of detailed regulation for most competition classes because the block is usually the largest mechanical component and limits the power through its capacity (swept volume), configuration, and general structure.

Material – Aluminum alloy is the best choice for low weight and thermal conductivity. This may be either die-cast or sand-cast. The grain structure of aluminum alloy engines can be improved by head treatment – this can include deep freezing for several days.

Capacity – Changing the bore and/or the stroke will alter the engine capacity. Usually there are competition limits on overboring, which will weaken the structure of the engine.

Configuration – This is the number and layout of cylinders. Now before you say that not much can be done about this point, it must be remembered

284 Motorcycle Engineering

that the first Cosworth V8 was made from two 4-cylinder engines put together, and Saab made a range of engines using the same components in different configurations.

Liners – These may be of the wet type or the dry type. The liners may be coated or plated to reduce friction or increase life. One common material is chrome – its shiny surface reduces friction and therefore can increase power output. The liners may also be made out of different materials; steel is normal for motorsport vehicles. If a cast iron cylinder block is being overbored – taken to a size above normal rebore limits – then it is often good practice to machine it out even more and then to add steel liners. This gives better durability and ensures a minimum cylinder wall thickness.

When new dry liners are fitted into a block, the fitting is usually an interference fit, that is, a press fit. This may cause irregularities on the

Figure 17.8 Hooded head lamp and shortened front guard, not to be used on rainy days.

inside of the cylinder. These are removed by honing with a very fine-grade oil stone. Of course, this procedure cannot be carried out with coated liners.

Deck height – The height of the top of the cylinder block in relation to the piston crown. The deck height affects the compression ratio. A minimum deck height of about 0.2 mm (0.010 in) (always check manufacturer's figures) is needed to prevent the piston crown from touching the cylinder head due to the dynamic forces at maximum engine speed.

Sealing – Increasing the engine's power output will put a greater pressure load on the engine's seals. Some engines are built without gaskets. As gaskets are used to allow for uneven surfaces, these engines are machined very accurately. A sealant is used between the mating surfaces.

Cylinder head gaskets may be replaced on race engines by ones made from different materials – frequently, this is one sheet of malleable metal. Where the compression ratios are very high, the block may be machined with annular rings around the cylinder bores to accommodate sealing rings.

The sump and other gaskets may be made from materials that are more able to cope with the high pressures and temperatures of competition engines – they will also be made to more exacting levels of accuracy than standard road versions.

Bearings – The crankshaft main and big-end journals are likely to be harder than those of road cars, so the bearing surfaces must be able to work with the materials. Normally the bearings are harder, containing more tin and less lead to make up the white metal alloy.

The bearing caps may also be upgraded by either fitted steel caps, which are then line bored to match the crankshaft, or fitted steel straps over the existing caps to increase their strength.

Pistons

For high-performance purposes, the piston is a very important part of the modification process.

As an overview, there are only a small number of companies that produce pistons in the UK and America, and notwithstanding the Asian market, they tend to produce for the worldwide motorsport range too. This is because piston production is very specialized precision engineering, requiring expensive machine tools and highly skilled staff. The motorsport piston supply is largely related to modification of standard pistons – although this is done at the manufacturing stage, not as a postproduction change. Or the application of a piston to one engine that is used as standard in another engine. If you think of the Ford range of engines that come in various capacities and are used in different models, you may get a better picture of this.

Table 17.2 Characteristics of Piston Design

No	Part of Piston	Design Characteristic	Reason	Comment
1	Crown	Height above gudgeon pin	Alter compression ratio and relation to deck height	
2	Crown	Shape	Changes to swirl and cut-outs for valve pockets	Valve pockets may need to be machined to cope with extra valve lift
3	Gudgeon Pin	Diameter	To cope with extra load	This will also involve modifications to the con rod
4	Gudgeon pin	Type of fit	To suit con rod	Types of fit include fully floating and interference
5	Skirt	Shape	Slipper skirt to reduce friction with cylinder wall and reduce weight	
6	Skirt	Design	Solid skirt to increase strength	
7	Skirt	Length	Short skirt to reduce friction with cylinder wall and reduce weight	
8	Rings	Number	Fewer rings – two or three – to reduce friction with cylinder wall	
9	Rings	Type	To enhance combustion sealing	Ring material may also be changed
10	Material	May be forged, billet, or cast from a variety of aluminum alloys	Will affect weight, strength, and machining processes	Pistons can be made in any size or design, but if nonstandard will cost accordingly

Figure 17.9 All set for Route 66.

Connecting rods

The most common modification to connecting rods (con rods) is balancing them. There are a number of different ways of balancing con rods. The basic principle is to ensure that they all weigh the same amount and that the weight of the top (little end) and the bottom (big end) is also equal on each con rod. This is done by supporting one end and weighing the other. To lighten the con rod, metal is moved by drilling holes in a lesser stressed section. Weight can be added by drilling holes and filling them with lead, which is much denser.

The con rods offer resistance to the engine, both from their mass – resistance to acceleration – and the air resistance moving inside the crankcase. The acceleration is improved by using lighter materials; in ascending order of cost, the options are aluminum alloy, titanium, and carbon fiber. The aerodynamics are improved by making them an aerodynamic shape.

Figure 17.10 Mudguards that work, leg shields, and comfortable saddle.

An economical improvement can be made to the original con rods by balancing, shot blasting smooth, then polishing.

If the stroke length is changed by changing the crankshaft – a very popular modification on many engines – then it may be necessary to change the con rod to ensure that the piston operates in the correct area of the cylinder bore.

When building engines, no matter which type of con rod is being used, you should:

- Check the oil way drilling
- Check the little-end fastening
- Check for straightness
- Check the big-end fitting, especially the cap and retaining nuts and studs or bolts

More big-end caps come loose than con rods break in the middle. Use the correct fastenings, and torque to the correct setting. As each piston and con rod is added to the block, check that the engine turns freely. Also ensure that the little-end fit is correct, as if it is not, the piston is likely to break at this point. That is the piston boss breaks away from the piston skirt, and the engine is likely to be wrecked.

On very high-performance engines the con rods should be seen as a service item that is replaced at each rebuild along with the bearings.

Crankshaft

Materials – The crankshaft is one of the main components in limiting the engine speed and power output. Standard crankshafts may be made from either cast iron (CI) or steel. The steel is likely to be a forged medium-carbon variety. For high-performance applications the steel is usually upgraded to one that will accept surface hardening readily. This usually means a slight increase in carbon content and the addition of other metals such as chromium, silicone, copper, and aluminum.

The two main types of surface hardening, designed to increase crankshaft life by reducing wear, are:

- **Nitriding** – Immersing in a *salt bath,* a solution of nitric acid (very hazardous) for a controlled (long) time period at a preset temperature. This is usually applied to more expensive steel alloys and increases the strength of the crankshaft.
- **Tuffriding** – A cheaper option for use with cheaper steels and only giving a surface hardness. This involves immersion of the crankshaft into a bath of (very hazardous) sodium cyanate at 570 °C for 2 hours.

Design – The design of the crankshaft has a number of factors for consideration when modifying the engine. Looking at the main ones:

- Configuration – This affects the firing order and balance; on six- and eight-cylinder engines there are a number of options.
- Webs and counterbalance – For high-speed engines, these are removed to make the crankshaft light; balance becomes dynamically controlled.
- Balance – Done on a balancing machine both statically and dynamically, usually in conjunction with the flywheel, con rods, and pistons.
- Thrust races – To prevent longitudinal movement, thin needle roller bearings may replace the plain metal ones to reduce friction.
- Oil ways – The oil passage ways are cross-drilled; that is, extra holes are drilled across the crankshaft to ensure a good oil supply to the big-end bearings.

Flywheel

As the purpose of the flywheel is to keep the engine turning between firing strokes, the faster the engine runs, the less the flywheel inertia is needed. So faster running engines and ones with more cylinders require flywheels with less mass (weight). The less the mass of the flywheel, the faster the engine will be able to accelerate. However, this will lead to uneven running at low engine speeds – not a problem on race bikes.

Balance – The flywheel will be balanced both statically and dynamically in conjunction with the crankshaft and other rotating parts.

Be aware – Changes to the flywheel may have an effect on dynamic balance, and it may also be necessary to look to the crankshaft and any counterbalance shaft mechanism.

Carburetor

Main modifications – The main modifications to carburetor arrangements are:

- Increasing the number of carburetors – ideally one per cylinder
- Increasing the size of the carburetors
- Changing the air filter (cleaner) for a high-flow one – for example, K&N
- Fitting air trumpets of different lengths
- Fitting water heated or nonheated inlet manifolds
- Altering the carburetor setting to give different mixture strengths under different running conditions

Injector systems

Fuel injection systems are typically modified in two ways:

- The electronically programmable read-only memory (EPROM) in the fueling ECU (Electronic Control Unit) is reprogrammed to give more fuel under particular conditions. This is called *chipping*, as the EPROM is a type of microchip. Chips can readily be bought for most popular sports bikes.
- To deliver the extra fuel, high-flow-rate injectors are used when the engine is chipped. These are sometimes identified by the color code green instead of the normal gray.

Throttle bodies

Instead of the normal injectors and manifold arrangement, each cylinder is fitted with short individual stubs which incorporate the injectors. This improves air and petrol flow.

Inlet manifold

These may be changed for ones that give different air flow characteristics and carburetor or injector arrangements.

Exhaust system

The exhaust system has the job of quietening the noise of the exhaust gas and cooling it before it leaves the vehicle. It will also, except on older motorcycles, incorporate a catalytic converter (cat) and a lambda sensor. Most modifications relate to improving the flow of the exhaust gas from the engine to the outside of the vehicle. Some are solely about altering the exhaust note to make it sound *sporty*. The exhaust note is very important to some people – it's about impressions of power. Typical exhaust modifications include:

- Change rear silencer for one with less restriction; chip it to suit.
- Change the manifold (called headers in the United States) for one that gives a better gas flow – usually this means longer individual pipes before the cylinders are joined.
- Complete system change – for serious performance increases, usually matching the inlet manifold and cylinder head and camshaft changes at the same time with a new chip.
- Change to a stainless steel exhaust system – most appropriate to classic bikes for longevity.

Supercharger/turbocharger

Superchargers and turbochargers are fitted to increase the amount of air and petrol going into the engine cylinder. The greater the amount of gas that goes into the cylinder, the greater the power output will be. Superchargers and turbochargers typically increase engine output by 30%.

The supercharger is in effect an air pump driven by the crankshaft – they started life on aircraft, where the air is less dense and needed to be compressed both for the engine and the passengers.

The turbocharger is driven by the exhaust gas; it has two turbines. One turbine is driven by the exhaust gas; the other turbine is driven by the shaft from the first turbine and compresses the air into the cylinder. On high-performance motorcycles, an intercooler may be used to cool the air between the turbocharger and the cylinders.

Although superchargers and turbochargers do the same job, they do it in very different ways.

Table 17.3 Comparison of Supercharger and Turbocharger

No.	Type	Drive	Characteristics	Advantages	Disadvantages	Comment
1	Supercharger	Belt from crankshaft	Straight-line pressure increase dependent on engine speed	Immediate response to throttle	Needs drive from crankshaft	Ideal for rapid acceleration – drag bikes
2	Turbocharger	Exhaust gas	Delayed pressure increase dependent on exhaust gas pressure and A/R ratio	Exhaust drive giving better economy for a given application	Delayed response to throttle – prompts use of twin turbochargers	Ideal for application combining power and economy on road bikes

Lubrication

The first modification to the lubrication system is to use a good-quality oil and oil filter. There are lots of alternatives for each type of engine and lots of different makes of each type. The supply of oil to the motor industry is both high value and high profit. So, some detailed care in the choice of oil is needed. Many manufacturers have a preferred brand, which is often one of their sponsors. For example, Ferrari recommends Shell and Shell sponsors Ferrari.

Oils are classified by the Society of Automotive Engineers (SAE, an institution based in America) based on viscosity (how they flow); the common grades are 10W-40 and 15W-40. The capability of a particular make of oil to do its job in an engine is classified by the American Petroleum Institute (API).

Actual mechanical modifications that are typically made to the lubrication system are:

Relief valve – The operating pressure of the relief valve may be changed; this may be by changing the spring, adjusting the spring setting, or replacing

Table 17.4 Examples of API Oil Classifications

No.	API Classification	Application	Comment
1	SJ	Motorcycles made in 2001 or before	
2	SL	Motorcycles made in 2004 or before	
3	SM	Motorcycles made in 2010 or before	
4	SN	Current – introduced in 2010	Current in 2020

the complete assembly. Usually increasing the oil pressure by about 10% will provide the ability to cope with greater bearing loads caused by engine power increases. Typically, oil pressure is 60 to 100 psi (4 to 7 bar).

Oil cooler – This is to control the oil temperature and prevent engine damage. Typical oil temperature is about 150 °C. The oil cooler must be fitted in a place where cold air flows freely – often this is just in front of the engine coolant (water) radiator. The flow of oil through the cooler may be controlled by a thermostat. The oil flow is usually in series with the oil filter.

Sump baffles – These prevent oil surge on corners, thus preventing oil starvation because the oil has moved away from the oil pick-up or oil burning because it is climbing the cylinder walls. Baffle plates may be welded inside an ordinary sump.

Oil pick-up – The modifications to this part include fitting a broader mesh filter to allow greater oil flow and moving the open end to the center of the sump to avoid oil starvation (see sump baffles).

Oil filter – A high-flow filter may be fitted to improve oil flow, or it may be remotely mounted for ease of changing and incorporation with the oil cooler.

Dry sump – Instead of the lubricating oil being in the sump underneath the crankcase, a separate oil tank is used. The oil from the engine components fall into the sump, but it is drawn out of the sump to the tank by a separate – scavenge – pump. The pressure pump draws oil from the tank to pressure-feed the engine's bearings.

VEHICLE-FIXED COORDINATE SYSTEM

The vehicle motions are defined with reference to a right-hand orthogonal coordinate system by SAE conventions. The coordinates originate at the center of gravity (CG) and travel with the vehicle. The coordinates are:

- x – forward and on the longitudinal plane of symmetry
- y – lateral out the right side of the vehicle
- z – downward with respect to the vehicle
- p – roll about the x-axis
- q – pitch about the y-axis
- r – yaw about the z-axis

EARTH-FIXED COORDINATE SYSTEM

Vehicle attitude and trajectory through the course of a maneuver are defined with respect to a right-handed orthogonal axis system fixed on the Earth. It is normally selected to coincide with the vehicle-fixed coordinate system at the point where the maneuver is started. The coordinates are:

X – forward travel
Y – travel to the right
Z – vertical travel; positive is downwards
Ψ – heading angle
ν – course angle
β – sideslip angle

With the fixed-vehicle system, you can log the bike at any point in time. However, if you are using this on a circuit, for instance, Brands Hatch, then it is helpful to have this information related to the circuit. You can then look at ways of improving your times for the circuit.

Figure 17.11 Lights to see where you are going and mirrors to see where you have been.

Figure 17.12 Very 1966 mod – yes, the personal plate is real.

Figure 17.13 Royal Airforce Blue.

Tuning and customizing 297

Figure 17.14 Incredible detail.

Figure 17.15 Keep-it-simple look – very effective.

Chapter 18

Inspection, test, and rebuild

The wider the range of motorcycles that you can inspect when training, the wider and more varied will be your experience and employability. Of course, you may choose to specialize at an early stage. Both approaches have advantages.

What is important is that you carry out motorcycle inspections and record your findings. Carry out inspection with the IMI (Institute of the Motor Industry) data sheets, which you can obtain direct from the IMI if you are registered. If you have not got access to these IMI sheets, then you may use MOT or service check sheets as guidance.

SAFETY FIRST

When approaching a vehicle for the first time, especially a damaged motor-sport vehicle, you must carry out a **dynamic risk assessment**. That is a

Figure 18.1 Deep socket.

Figure 18.2 Socket drive adaptors.

mental assessment of the situation. If you are on duty at an event, maybe as a team mechanic or a marshal, you will be first on the scene. In this case you need to consider the following:

- Is it safe to get near the vehicle – think about location, other traffic, and other people; you must put your safety as the first priority
- The next priorities are making the scene safe, calling for help, and the application of first aid and perhaps calling the paramedics

Table 18.1 Using Your Senses

Sense	Checks	Cautions
Sight	How does this vehicle look? Is it level and square? Are there any leaks or stains? Signs of damage or misuse?	Use eye protection
Sound	How does this vehicle sound? Listen to the different systems or parts.	Use hearing protectors
Smell	Is there a smell that might indicate a leakage or overheating?	Wear a mask
Taste		Not advised
Touch	Use your fingertips to check for damage or wear. Use a nail to check whether a blemish is raised or sunken.	Use hand protection
Kinesthetic	Feel the operation of controls or mechanical linkages for smoothness.	Be prepared for the unexpected

Figure 18.3 Combination spanner set.

If you are approaching the vehicle after its recovery, even when it is in the pits or workshop, a dynamic risk assessment is needed.

Some typical examples are:

- Petrol spillage
- Hot engine and brake parts
- Jagged edges of bodywork

Use your senses

When inspecting a motorcycle, it is always good to use your senses, but do it with care.

Respect for motorcycles

As a technician YOU ARE RESPONSIBLE for the vehicle that you are inspecting; therefore, you must NOT cause damage to the vehicle, even if the vehicle that you are inspecting is seriously damaged. You never know what repairs or salvage may take place. You are expected at all times to:

- Protect from bad weather
- Jack-up and support the vehicle safely using appropriate jacking points
- Ensure that the systems are treated with care
- Remove finger marks

Table 18.2 Tire Markings and Wheel Serviceability

Task	Detail	Action points	Result
Use necessary PPE	Mechanic's gloves	Look out for hot tires and sharp edges	
Correctly identify relevant tire data for vehicle	Size, pressure, type, fitment	Use either vehicle mfg or tire mfg data sheets	
Locate and identify tire markings	Tire and wheel size, speed rating, load index, tread wear indicator, aspect ratio	Sketch tread wear indicator between tread ribs	
Measure tread depth	Use MOT green depth gauge	Check across full width of tread and all circumference	
Examine tire condition	Look for damage to the tire	Check the tire pressure	
Examine tread wear patterns	Look for uneven wear (see next), skid wear and patch wear on tread		
Identify reasons for abnormal wear patterns	State causes of edge or wear at one point		
Examine wheel condition	Look for damage to wheel		
Check valve condition and alignment	Is it in straight?		
Record faults		List wheel and tire faults	
Complete data collection sheet		Record findings on data collection sheet	

The following tables are designed to give you a practical guide to inspecting motorcycles. You may choose to use them as a guide and to make up your own checklists.

TEST AND REBUILD

Before carrying out any overhaul work, you should check the following:

- The customer fully understands the cost implications
- You have access to any special tools or equipment that is needed
- Spare parts are available
- Repair, setup, and test data are available

Inspection, test, and rebuild 303

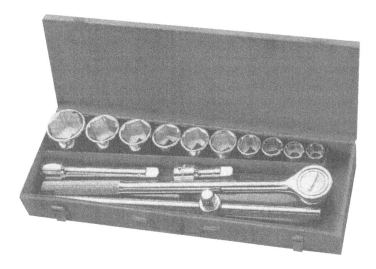

Figure 18.4 Socket set.

Figure 18.5 Stud remover.

Table 18.3 Pre-event Setup – Workshop

No.	Task	Detail	Typical Answers	Action Points
1	Use necessary PPE	Mechanic's gloves, goggles		
2	Obtain vehicle setup data	Check with appropriate authority	Use setup management system	This may need a signature
3	Check previous data and driver comments		Use setup data log	
4	Check battery serviceability	Use battery test procedure		System will not operate without a sound power supply
5	Check oil and fluid levels	Note variance		
6	Check and adjust suspension and corner weights	Refer to data	Complete setup documents showing changes	
7	Check and adjust steering geometry	Castor, camber, SAI, toe-out on turns	Complete setup documents showing changes	
8	Check brake conditions	Measure pads, discs	Record on inspection sheet	
9	Check driver safety equipment			
10	Carry out spanner check	Verify against checklist; rectify as needed	Note any loose, damaged, or broken fastenings	
11	Record faults		Use setup data file	Record any system faults
12	Complete data collection sheet			Record findings on data collection sheet

COSTS

The cost of overhauling a component – such as a gearbox – can, in some cases, be greater than that of a new one. On the other hand, given that not many parts require replacement, the overhaul may be one-tenth of the new item. The biggest factor is often the labor cost followed by the cost of obtaining the parts needed. So, the labor charge-out rate is a very big factor.

Let's look at a possible example using a labor rate of £85 per hour.

In addition, the replacement gearbox will have a period of warranty attached, giving the customer peace of mind.

Inspection, test, and rebuild 305

Figure 18.6 Aircraft spanners.

When it comes to replacement parts such as engines and gearboxes, the customer may wish to have the original part overhauled rather than fit a replacement simply to maintain the originality of the vehicle. Rather like the blacksmith who said, "This is the original hammer that I've had since I was an apprentice – it has had five new heads and six new shafts."

Beware, however, when buying or selling units that there are many different definitions or uses in this field; for example:

Original Parts – OEM (original equipment manufacturer) – The units used on the vehicle when it was new. These are often not actually manufactured by the vehicle manufacturer but sourced from an approved supplier. The boxes may have the logo of the vehicle manufacturer to show approval, but exactly the same item without the logo on the box may be obtainable from the local motor parts factors for 50% of the original price. Also, be aware that many manufacturers – particularly German ones – often use recycled parts or products made from recycled materials; such parts are still, by definition, OEM and will carry a price premium.

Remanufactured units – This applies to engines and gearboxes in particular. The components are completely stripped, inspected, measured, and crack tested and then all the machined faces are remachined and new wearing parts fitted, for example, bearings and pistons.

Overhauled units – In this case the units are stripped, cleaned, and inspected, but only the parts that are outside tolerance are replaced.

Pattern parts – These are look-alike items that fit exactly and do the same job as the OEM part but are made by another company. Sometimes

Table 18.4 Pre-event Inspection – At Event

No.	Task	Detail	Typical Answers	Action Points
1	Use necessary PPE	Mechanic's gloves, goggles		
2	Check previous data and driver comments		Use set-up data log	Check previous data and driver comments
3	Check tires	Fitment (direction/position), pressure, type, compound	Use setup data log	
4	Check oil and fluid levels	Note variance	Check oil and fluid levels	Note variance
5	Check and adjust steering		Complete setup documents showing changes	
6	Check security of body work	Panel attached, clips closed, screen clear, decal correctly located	Use checklist	May choose to photograph for record
7	Check seat belt harness with driver in place	May need adjusting		Note any presets
8	Arm fire extinguisher system	Check setting, switches		
9	Record faults		Use data log	Record any system faults
10	Complete data collection sheet			Record findings on data collection sheet

the pattern parts perform better than the originals, and often they are made by the same company that makes the originals but sold under a different brand name. Usually these units are 30% cheaper than the OEM equivalent.

HEALTH AND SAFETY AND THE ENVIRONMENT

Before carrying out any work whatsoever, you must ensure that you fully understand the health and safety issues and are able to comply with them along with the relevant environmental regulations and codes of practice. See the separate chapter on this.

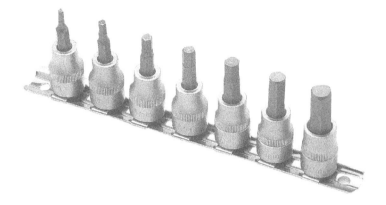

Figure 18.7 Hexagonal internal sockets.

Workshop

To be able to carry out any form of overhaul or repair – apart from the running repairs during competition – some form of workshop is needed. It sounds obvious, but often the functions of the workshop are forgotten about. Let's have a look at some of the reasons for a workshop:

- Provide safety, security, and protection from the weather for the motorcycles
- Provide safety, security, and protection from the weather for the engineers and other staff
- Provide safe and secure storage for the tools, equipment, and spares
- Provide somewhere to work on the motorcycle
- Provide somewhere to overhaul and repair the units and components
- Provide an office facility

Workshops come in all sorts of shapes and sizes. I have worked in them varying between a lock-up garage in a block without a permanent power supply and a Formula 1 team in a three-story futuristic building with an underground entrance using solar power and heat pumps to give a zero-carbon footprint. The ideal workshop is one where the motorcycles can flow through in one direction. This should fit in with the loading and unloading of the machines from the transporter. Some form of docking station is preferable so that the machine is kept both secure and dry on both its outwards and inwards journeys.

The race bike workshop is much more than a garage in the fact that it has a wide range of activities and a larger number of departments. One company I was involved with included a museum, car sales, and mail order. We are not discussing these areas in this chapter. If you are employed in the motorsport industry, you may be involved in all aspects of motorsport

Table 18.5 At-Event Inspection

No.	Task	Detail	Typical Answers	Action Points
1	Use necessary PPE	Mechanic's gloves, goggles		Remember components and systems remain hot for a while at the end of a stage or round
2	Check new data and driver comments	Download data logger and record verbal comments	Use setup data log	Check previous data and driver comments
3	Raise vehicle and remove wheels	Visual inspection		
4	Inspect wheels and tires for wear/damage		Record wear findings	
5	Carry out spanner check	Verify against checklist; rectify as needed	Note any loose, damaged, or broken fastenings	
6	Refit wheels	Check security and bearings		
7	Check fluid levels and pressures		Complete check sheet	
8	Check security of body work	Panel attached, clips closed, screen clear, decal correctly located	Use checklist	May choose to photograph for record
9	Refuel	Follow team procedure	Record fuel taken	Remember that petrol is highly flammable
10	Record faults		Save data file	Record any system faults
11	Complete data collection sheet			Record findings on data collection sheet

vehicle overhaul or just one tiny part. No matter how small, it will be an important function, and it will help if you have the bigger picture of what goes on in other departments. Some motorsport companies do just one of the functions covered in the next few pages.

Cleaning-down area – When entering into the building there will be a cleaning or wash area. Depending on the sort of competition involved, this will vary. Obviously, an off-road machine will have more mud to remove than a road machine – unless maybe the rider has had an off. This area may also be used for removing body panels – especially damaged ones – and

Inspection, test, and rebuild 309

Figure 18.8 Universal puller set.

Table 18.6 Post-Event Inspection

No.	Task	Detail	Typical Answers	Action Points
1	Use necessary PPE	Mechanic's gloves, goggles		Accurate recording and signing-off against names is recommended
2	Obtain and collate race data	Set up sheets, data logging, fuel and tire records		There may be both electronic and paper logging systems
3	Analyze race data	Record comments and make to-do list	Jobs to be completed before next event	
4	Check and correct log records	Team debrief and post-event analysis	Include data check and other comments	
5	Clean and secure for transporting	Check for damage and fit transportation tires	Photograph for record	
6	Drain fuel if appropriate	Use save storage procedure		Measure fuel left for fuel consumption calculations
7	Remove and charge battery			Gel batteries must be slowly charged
8	Complete data records	Data logging and vehicle records	Complete all data logging and setup sheets	Secure storage of data is essential – back up all files and keep in secure place
9	Complete data collection sheet			Record findings on data collection sheet

Table 18.7 Gearbox Overhaul Cost Comparison

Overhaul Costs		Replacement Costs	
Item	£	Item	£
Remove and refit gearbox 1.5 hours	127.50	Remove and refit gearbox 1.5 hours	127.50
Overhaul gearbox 5 hours	425	Replacement gearbox	500
Parts and oil	175	Oil	15
VAT	127.31	VAT	112.43
Total	854.81	Total	754.93

other items such as underpan guards and changing the wheels to ones more suited for the workshop.

Environmental Protection Act (EPA) and associated Local Authority (LA) Regulations and Building Regulations (Building Regs) require that all buildings comply with certain criteria. The cleaning-down area must have a drainage trap so that contaminated water, mainly meaning contaminated by oil and chemicals, is not allowed to enter the main drainage system. It is normal to recycle the cleaning-down water using a filtration and treatment system. This is environmentally clean and saves on the water bill.

Preparation area – An area of bays where each machine can be worked on individually. The layout of the bays and the equipment in them will depend on the type of machine. It is normal to have ramps, stands, or

Figure 18.9 Small socket set.

Inspection, test, and rebuild 311

Figure 18.10 Torx socket set.

trestles so that the vehicle is at an ergonomically sound working height. For a racing team it is usual practice to have a designated bay for each machine and a designated technician too. In this case the bay will be kitted out with that technician's tools and the equipment and spares appropriate to the vehicle. In a jobbing shop – where motorcycles are prepared for a variety of owners and/or drivers – the bay will be the sanctity of the technician, with motorcycles moving in and out as needed.

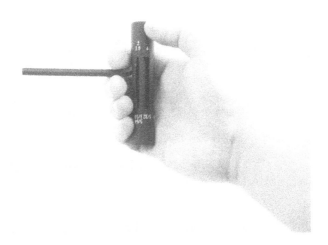

Figure 18.11 Allen key with built-in handle.

In the case of major unit overhaul, the bay may be used to remove the unit only and then the machine moved to a compound area for secure storage until the unit is ready to be reinstalled.

Machine shop – This is where the machine tools and similar items are laid out and used. Competition and custom machines tend to use a number of bespoke parts – ones made especially for that vehicle. So, the machine shop is needed for both manufacture and overhaul. Also, this may be where a number of special tools are kept – the ones which require floor mounting as opposed to those that can be taken to the benches in the preparation bays. Typically, a machine shop may contain some or all of the tools in Table 8.8.

Table 18.8 Machine Shop Equipment

No.	Item	Specification	Purpose	Comment
	Small lathe	6- to 8-in (150 to 200 mm) swing with 18 to 24 in (450 to 600 mm) between centers	Making small items such as spacers and cleaning up round parts	This will need a range of tools and chucks
	Off-hand grinder – small	Approximately 6-in (150-mm) diameter wheels – one fine, one course	General sharpening and cleaning	
	Off-hand grinder – large	Approximately 10-in (250 mm) diameter wheels	Sharpening drills and tools	Keep flat for accurate work
	Pillar drill	5/8-in (15-mm) chuck and variable speeds	Variety of drilling	Need variety of vices, or clamps, and drills and countersinks
	Band saw	Approximately 4- to 6-in (100- to 150-mm) cut	Cutting up steel stock	
	Hydraulic press	20 ton	Removing and replacing bearings and pins	
	Milling machine	Five-axis CNC milling center or manual type	Manufacturing small parts	Used in conjunction with CAD system
	Buffing/polishing	Floor-mounted buffing and polishing heads	Finishing parts	
	Parts cleaning bath	Chemical cleaning bath with pressure spray	Cleaning parts	

Bench work area – The bench work area is often around the outside of the work bays, using metal-topped benches with drawers and cupboards underneath. Vices and other tools may be mounted on the benches.

Where the work is solely unit based – such as overhauling engines and gearboxes – the benches may be aligned in rows separate from the motorcycles, with the use of stands or rigs for the gearboxes or other major components. The drawers or open racks will then contain the special tools needed for the job in hand.

Fabrication area – This is where items are fabricated and welded. Usually contains rollers, bending machine, croppers, and MIG or TIG welding equipment. Specially trained engineers in this area will provide these services to enable you to carry out your overhaul and repair tasks.

Composites shop – Where composite components are manufactured or repaired – a specialist clean area. This shop is staffed by specialist technicians who will make new parts or carry out specialist repairs to enable you to overhaul, modify, or carry out custom jobs.

Design studio – Where vehicles and components are designed and modified, usually using computer-aided engineering (CAE). The computer-aided design (CAD) is connected to the computer-aided manufacture (CAM) machine tools such as the five-axis milling center. You will find the design staff supportive in providing technical data for your overhaul procedures.

Model shop – Named because they make models, or maquettes, of vehicles for design and testing purposes, including scale models for the wind tunnel, as well as specialist full-scale parts such as airfoil sections. The model shop is a source of both data and specialist parts and skills. The model shop in larger or older firms may incorporate clay and/or wood handling equipment and skills.

Paint shop – Where the vehicles are painted. Again, this shop has specialist staff and equipment. The overhauled machine, or its panels, will be refinished in this shop. An interesting move in refinishing is to use vinyl film instead of paint. For example, the U.S. Army dragster, which is probably the fastest race car ever made, uses vinyl film. This was applied in their workshop at the Indianapolis Raceway, known as the *Brickyard*. The comment of the technician applying the material, which was printed and cut on site, was, "It's lighter than paint." When it is on, it is almost impossible to tell that it is not paint.

Parts and storage – The safe and secure storage of parts, both new ones and ones waiting for completion of the overhaul, are very important. They can easily go missing. Small parts have an attraction for the floor (it's called gravity – 9.81 m/s^2); then they roll behind the largest possible box so that you can't find them. That aside, competition motorcycle parts cost 10 or 100 times the equivalent of the road-going vehicle. And some people just want to have the damaged piston out of the number 12 car that didn't win the last race of the season or something similar.

Let's have a look at this in a bit more detail. You need a secure storage area large enough to store the large parts and a set of drawers and trays for storing the smaller parts. When you are stripping a component – such as a gearbox –you need a tray system laid out to keep all the small parts – nuts, bolts, washers, spacers and so on – in order so that they can be assembled in the reverse order. If you are doing repetitive work – such as overhauling gearboxes – you will probably be provided with suitably marked-out trays; you will also be able to identify each component without any thought.

To ensure absolute recognition, a number of procedures are used; two frequently used ones are:

- Plastic bags, like food storage bags – Put each component into a bag and label the bag with the part name, part number, customer details (motorcycle number or VIN), and any other details – such as *left rear*. The bags are then placed in a tray, which also has the customer, motorcycle, and unit identification.
- Photographs – Take photographs of both the assembled and the stripped parts. Record the photograph numbers (most digital cameras do this automatically), and reference the numbers to your notes.

These two techniques are used extensively in specialist firms, such as when repairing or overhauling high-performance, high-value, and custom motorcycles. As it is unlikely that a workshop manual exists, you'll rely on the parts you have, so look after them, make notes, and take photographs. Motorcycles like Ducati come in lots of variations for the same model range; owners frequently change seats, engines, and gearboxes. Hayabusa owners frequently change the rear panels and the rear suspension.

You will also need a storage system for everyday consumables, that is, items that are used on a day-to-day basis in the repair or overhaul of vehicles or units. Such items may include:

- Cleaning cloths
- Polishes and detergents
- Hand cleaning materials
- Specialist cleaning solutions
- Screws, nuts, and bolts
- Specialist fixings – such as toggle fasters
- Washers, spacers, and shims
- Locking wire
- Wiring cable
- Electrical connectors and fasteners
- Tape and adhesives
- Gaskets and seals

Inspection, test, and rebuild 315

Figure 18.12 Locking wire.

These items may be charged out against jobs in two ways:

1. Individually against the job number
2. On each customer's bill as either a percentage of the bill or prorated against the number of hours worked

Dynamometer and test shop – If working on complete motorcycles, you will need to have access to a rolling road dynamometer – referred to as the dyno or the rolling road by most staff. This allows the rear wheel to sit on the dyno rollers and the power and torque measurements to be taken. Running any competition engine makes a lot of noise, so this is usually situated separate from the main building, or suitably noise insulated from it.

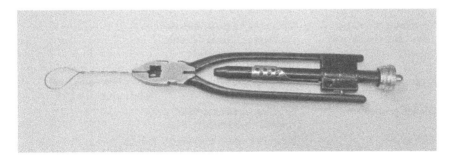

Figure 18.13 Locking wire tool.

Table 18.9 Overhaul Equipment

Item	Purpose	Note
Motorcycle lifting tables	Raise motorcycle	
Paddock stands	Keeps machine level	Front and rear wheel versions, with adaptors
Body scanner	Scanning bodywork to produce CAD drawings	Good when rebuilding historic machines
Cam profiler	Regrinding camshafts	Make cam profiles to your design
Wheel alignment level	Checking steering geometry and wheel position	
CMM machine	Accurately measuring components for CAD drawings or reverse engineering	
Coil spring gauge	Testing coil spring rate	Check all springs
Compression gauge	Testing engine compression pressure	Compare reading of each cylinder
Scales	Checking weight distribution front to rear	Adjust suspension and redistribute weight as needed
Crankshaft grinder	Regrinding crankshafts	
Dial test indicator (dial gauge)	Measuring movement – such as for valve lift	
Durometer	Measuring the hardness of tire treads	Check temperature first
Engine boring equipment	Reboring cylinder blocks	
Granite table	Providing a smooth and level surface on which to set up the vehicle	Cost is into £millions, used for world-class vehicles
Horizontal milling machine	Milling surfaces – such as cylinder head faces	
Laser or light suspension aligning gauges	Checking steering and suspension alignment	The manual system can achieve the same results
Mercer gauge	Measuring the diameter of a cylinder bore	
Micrometers – range, internal and external	Measuring inside or outside surfaces – such as cylinder bores and crankshaft bearings	
Pressure washer	Cleaning mud of the vehicle	Ensure EPA compliance when used
Surface grinder	Grinding surfaces such as cylinder head faces	
Turn tables	Measuring wheel movement	Used in conjunction with wheel alignment level

(Continued)

Table 18.9 (Continued)

Item	Purpose	Note
Tire machine	Removing and refitting tires	Special machine needed with aluminum alloy rims
Tire pressure gauge	Measuring tire pressures	Need accurate gauge on motorsport vehicles – check temperature before adjusting
Tire temperature gauge	Measuring tire temperature	Take three measurements on each tire: inside, middle, and outside of tread
Vertical milling machine	Milling tasks such as when enlarging inlet ports	
Welding equipment (MIG or TIG)	All kinds of joints and repairs	
Wheel balancer	Static and dynamic wheel balancing off the vehicle	Use only approved weights – usually stick-on inside of the rim

If just the engine is being tested, a test cell is used. A test cell, of course, requires much less space than a rolling road. This takes the form of the engine mounted on a frame, which is attached to the dynamometer (dyno). Because of their exceptional noise, aircraft engines are tested underground; some motorcycle race teams do this too.

Figure 18.14 Martin Penning movable repair pod.

Checking and loading area – This is a secure area for the prepared motorcycle to be kept safely prior to and during loading onto the transporter.

Office – Office space is needed for a number of functions in motorsport; these include:

- Reception – To meet and greet customers and to record customer information and carry out tracking of customer repairs.
- Finance and administration – To control the flow of cash and communications with customers and suppliers.
- Meeting area for the team and directors – Often this takes the form of a boardroom with a big table and chairs that can be used for a number of functions.
- Engineering office area – With a racing team, the race engineers will need an area to work on data from testing and racing to analyze the data and work out strategies for future developments. This usually takes the form of bar-type desks, where the laptop computer can be docked into the team network – usually still hard wired for security, though high-level encryption may be used with a suitable wireless virtual private network (VPN). This is a separate area from that of the design team, who are working on upgrades and new designs for future vehicles and components.

In addition to the functional part of the activities in the office area, this area can be used for displaying trophies – *silver ware* – and photographs. Therefore, the security of the office must be considered with appropriate locks and an alarm system. Interestingly many of the WSB teams have private museums at their headquarter (HQ) offices.

Tools and equipment

Table 18.9 lists the equipment needed for the workshop. In this section we will look at the fuller range of equipment that is used in motorcycle and unit overhaul.

Data

Sources of data are very important. The obtaining and storage of data is very much a profession in itself. Think of the word library – it does not just refer to books, but all forms of stored information that include photographs, films, tapes, posters, CDs, DVDs, and other electronic storage media.

As a motorcycle technician, you will have to collect, collate, store, use, and communicate data.

There is a saying that knowledge, another word for data, is power. If that data means 0.1 of a second off each lap, that is very powerful, and therefore

Inspection, test, and rebuild 319

Table 18.10 Data Sources

No.	Source	Typical Content	Accuracy Limits	Comment
1	Workshop manual	Technical service data	Detailed and checked	
2	Autotrader	Vehicle sales		
3	Competition vehicle log	Setup data for vehicle, work done, parts fitted	If correctly maintained	Must be maintained in detail to ensure continuity of operation and to save time
4	Customer record/file	Name, address, vehicles owned, contact details	If correctly maintained	Highly confidential
5	Data logger	Vehicle operation data including acceleration, braking, LAG, gear, throttle position	Digital data – as good as system will allow	PI
6	ECU data	Codes for system operation setup and fault codes	Digital data	Each system may have a separate ECU – a reader will be needed to access data
7	Glasses guide	Vehicle model guide	Detailed and checked	
8	Haynes (or similar) workshop manual	Vehicle and units service and repair procedures	Detailed and checked	Haynes offers an excellent range of manuals at competitive prices
9	Lap/section time from circuit system	Speed and timing of laps and sections	Digital data – as good as instruments will allow	Data available from race control
10	Manufacturer's workshop manual, parts manual	Vehicle and units service and repair procedures	Detailed and checked	Often to be read in conjunction with training materials
11	Parker's price guide	Vehicle identification and pricing	Detailed and checked	
12	Stack system	Speed and timing of laps and sections	Digital data – made to high level of accuracy	In motorcycle timing system – very useful for testing
13	Vehicle service book	Records of service and mileage	If correctly maintained	
14	Your company records	Vehicle changes and modifications	If correctly maintained	Annotated

Table 18.11 Pre-overhaul Data Check

No.	Data	Source	Comment
1	Test data	Report on test	This should be added to vehicle log
2	Unit removal data	Workshop manual	
3	Unit test data	Unit manufacturer	
4	Unit stripping data	Unit manufacturer	
5	Equipment operation data	Equipment manufacturer	
6	Replacement parts	Parts manual	Parts used should be recorded
7	Unit assembly and test data	Unit manufacturer	This should be added to vehicle log
8	Setup data	Vehicle setup log and/or unit manufacturer	This should be added to vehicle log

very crucial and valuable, data. As a technician obtaining, using, and storing data are high priorities.

Most manuals and other commercial sources of information are available on CD/DVD and online through a subscription agreement, as well as in paper form.

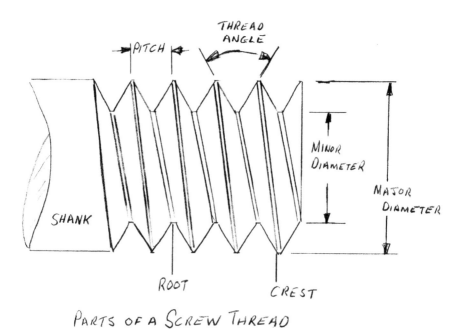

Figure 18.15 Parts of screw thread.

Before you carry out any overhaul work, you will need to check that the data are available.

Parts and fixings

On racing and competition vehicles of all types, fixings are very important. It is good when working on a motorcycle to remember a bit of engineering science for the formulas. To make these formulas more graphic for you to imagine, thinking of them in these different ways may help:

1. Force – If you want to fix a post into the ground, take a big sledgehammer (14 lb/6.3 kg mass) and swing it quickly, holding the end of the shaft for rapid acceleration; now in terms of a motorcycle, it is the weight (mass) multiplied by how fast it is accelerating. If you move the formula round, you can see that to make it accelerate faster, you can either fit a more powerful engine – increasing the force – or make it lighter.
2. Stress – As you can see, this is inversely proportional to the area taking the stress – so usually the smaller the part, the more stressed it will become.

Bearing these factors in mind – if we make the vehicle lighter and faster, we tend to make the parts more stressed, so we need to be sure that the parts used will cope with the situation. Therefore, the parts and fixings tend to be made out of materials that are both light and strong. Typically these are high-strength steel (HSS), aircraft-grade aluminum alloy (7001 or similar series), and titanium. These materials are much more expensive than those used in standard motorcycles; also the parts are made in smaller numbers and are therefore much more expensive to produce.

Aluminum alloy and titanium are both easily damaged, and their appearance can soon be marked, so ensure that they are handled with care and that the correct tools are used when working on them.

Procedures

When carrying out any overhaul work, you must follow the procedures set out in the appropriate manuals and data sheets. Technical explanations for some tasks are covered in the appropriate chapters.

In all cases follow these basic steps:

- Obtain the necessary data
- Clean the vehicle in the vicinity of the unit to be removed
- In a clean work area remove the unit to be overhauled, following the appropriate safety sequences
- Take the unit to the bench, or mount on a stand

- Strip the unit, noting the position of parts and using appropriate parts storage trays
- Repair and rebuild as needed
- Replace the unit and set up as per data
- Thoroughly test and recheck setting against data

Products

The use of products – oils, greases, and other chemicals – in the motor industry is big business. The correct use of products can often save time and money for the customer and increase sales revenue for you.

When you have carried out an overhaul task, you will probably need some form of product for lubrication or cooling. The customer should be told of this and, if possible, given an after-care leaflet or card with information about the product and its use to prolong the life of the overhauled unit.

EXAMPLE OF AFTER-CARE LEAFLET

BROOKLANDS GREEN – BRAKE CALIPER CARE

1. Your brake calipers have been overhauled and should give you a high standard of braking.
2. You are advised to use *AP Racing DOT 5.1* brake fluid, which complies with SAE J1703 – this is ideal for high-performance road, competition, and track day use.
3. To prevent corrosion the brake fluid should be **changed frequently** – we recommend at least each racing season.
4. To prevent brake squeal use a **copper-based grease** between the pad back plate and the caliper assembly.
5. Be sure to **fully vent** (bleed) your brakes after fitting, before driving.

Clusters – Motorsports Valley

The implications of parts supply, knowledge and data, tools and equipment, and transportation all revolve around cost. In motorsport there tends to be a team working spirit – yes, even between competing teams. After all, there would be no motorsport if there weren't any other teams. So, rather paradoxically, although it is all out on the track, you need to work with your competitors to achieve your aims of winning. Because of this companies tend to work in clusters – small groups based at circuits, in trading estates, or sites with good linkages to each other. In the UK, the bulk of the major F1 companies is located in Motorsport Valley. This is geographically following the Thames Valley from Oxfordshire, through the Home Counties, down to Kent. The communication routes use the M4 and M40 leading into the M25 and exiting with the M2.

The motorcycle industry followed from the bicycle industry; their home was originally in Coventry. Royal Enfield has a facility in Redditch and Triumph has one in Hinckley, both on the outskirts of Coventry, and both are conveniently located near Donnington Park Circuit, where a lot of motorcycle racing takes place.

Networking between the motorsport companies and the competitors is common. Many companies solely exist to be able to race, while others operate in other branches of performance engineering, such as aerospace and super yachts, to provide an extra income stream. Most networking place in the Motorsport Valley. Organizations like the Motorsport Industry Association (MIA), the Institute of the Motor Industry (I M I), the Institution of Mechanical Engineers (IMechE), the Motorsport Institute (MI), and Autosport Magazine organize some of these events. Often, they are funded by government bodies such as UKTI and local development agencies. In the United States, the magazine and show organizer Performance Racing Industry (PRI) is one of the main networking organizations for the industry; they have links with the British institutions as well as the Society of Automotive Engineers (SAE).

Chapter 19
Sustainability

As an engineer you have a duty to look after our planet and its surroundings. The Engineering Council and the Royal Academy of Engineering produced this statement as part of the ethical principles that engineers should follow:

Engineering professionals have a duty to obey all applicable laws and regulations and give due weight to facts, published standards and guidance and the wider public interest. They should:

- *hold paramount the health and safety of others and draw attention to hazards*
- *ensure their work is lawful and justified*
- *recognise the importance of physical and cyber security and data protection*
- *respect and protect personal information and intellectual property*
- *protect, and where possible improve, the quality of built and natural environments*
- *maximise the public good and minimise both actual and potential adverse effects for their own and succeeding generations*
- *take due account of the limited availability of natural resources*
- *uphold the reputation and standing of the profession*
- *Further details can be found at www.engc.org.uk*

THE THREE CENTRIC CONCERNS OF SUSTAINABILITY

Sustainability is a very current term, but often misused and misquoted. There is no clear definition, nor an absolute one, although others may not agree. What we are trying to do with sustainability is to ensure that our planet lasts into the future. We don't know what the future will be like – that is one of the definitions of the future. It is also part of the theory of quantum physics. However, we do know that we want it to last and we

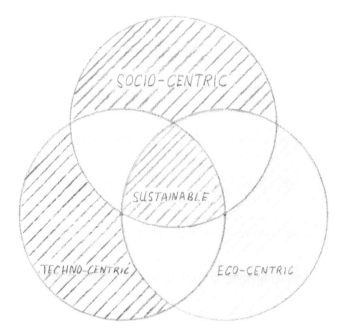

Figure 19.1 Three centrics of sustainability. Something is only sustainable when social, technical, and ecology concerns overlap.

want to get the best value out of it. Of course, value is another word with lots of different meanings for different people. The basic premise of sustainability is to ensure that we have convergence, or overlap, between the three centrics. The centrics are areas of concern about what we are doing with a view that whatever it is that we are doing will work properly and last. This is almost another way of looking at the Engineering Council statement of ethics.

Techno-centric concerns are about will it work properly, can we trust the science and technology to deliver what it says? There will always be Luddites who oppose new technology. If you look at the history of toll roads (see the timeline) and the resistance to the use of any other form of transport than horses.

Tech Note

There's a large plaque on the wall of the Ford factory in Essex that reads *If I'd asked people what they wanted, they would have said faster horses.* – Henry Ford

---

Socio-centric concerns are about social acceptance – getting people to accept the changes and developments. There will always be the

NIMBYs – not in my backyard people. As an engineering lecturer I teach apprentice engineers from the electrical power supply companies. One year a group was replacing switching gear in a substation; this, of course, meant disruption in the local area. It led to lots of complaints – the residents could not understand that this was to prevent them from having power cuts for the next 20 years. Nobody said thank you when it was done, even though they'd been informed all about it. My apprentices got their degrees and won't be back there until they are ready to retire.

Eco-centric is about the ecology. That is the plants, the animals, and the insects. The study of this is called natural history. There are libraries written about natural history but still very little is known. Even less is known about the effects of changes in the environment on the ecology system or the food supply chain.

TRANSFORMING OUR WORLD: THE 2030 AGENDA FOR SUSTAINABLE DEVELOPMENT

The United Nations have an agenda for sustainable development. It was set in 2015 for a period of 15 years – the length of a generation – to go to 2030. This is the introduction.

Preamble

This Agenda is a plan of action for people, planet and prosperity. It also seeks to strengthen universal peace in larger freedom. We recognise that eradicating poverty in all its forms and dimensions, including extreme poverty, is the greatest global challenge and an indispensable requirement for sustainable development. All countries and all stakeholders, acting in collaborative partnership, will implement this plan. We are resolved to free the human race from the tyranny of poverty and want and to heal and secure our planet. We are determined to take the bold and transformative steps which are urgently needed to shift the world onto a sustainable and resilient path. As we embark on this collective journey, we pledge that no one will be left behind. The 17 Sustainable Development Goals and 169 targets which we are announcing today demonstrate the scale and ambition of this new universal Agenda. They seek to build on the Millennium Development Goals and complete what these did not achieve. They seek to realize the human rights of all and to achieve gender equality and the empowerment of all women and girls. They are integrated and indivisible and balance the three dimensions of sustainable development: the economic, social and environmental.

The Goals and targets will stimulate action over the next fifteen years in areas of critical importance for humanity and the planet:

People

We are determined to end poverty and hunger, in all their forms and dimensions, and to ensure that all human beings can fulfil their potential in dignity and equality and in a healthy environment.

Planet

We are determined to protect the planet from degradation, including through sustainable consumption and production, sustainably managing its natural resources and taking urgent action on climate change, so that it can support the needs of the present and future generations.

Prosperity

We are determined to ensure that all human beings can enjoy prosperous and fulfilling lives and that economic, social and technological progress occurs in harmony with nature.

Peace

We are determined to foster peaceful, just and inclusive societies which are free from fear and violence. There can be no sustainable development without peace and no peace without sustainable development.

Partnership

We are determined to mobilize the means required to implement this Agenda through a revitalised Global Partnership for Sustainable Development, based on a spirit of strengthened global solidarity, focussed in particular on the needs of the poorest and most vulnerable and with the participation of all countries, all stakeholders and all people.

The interlinkages and integrated nature of the Sustainable Development Goals are of crucial importance in ensuring that the purpose of the new Agenda is realised. If we realize our ambitions across the full extent of the Agenda, the lives of all will be profoundly improved and our world will be transformed for the better.

The full document can be read at: https://sustainabledevelopment.un.org/post2015/transformingourworld

DISCUSSION

There is no simple formula, nor standard working procedure, for sustainability. Having given a brief introduction to some of the topics of sustainability, the author suggests that the reader considers sustainability when appropriate.

All engineers need to be aware of the Engineering Council statement to work in a professional manner. All engineers also need to be aware of the concept of sustainability and the three centrics. You might discuss the following topics with colleagues:

1. If you bought a motorcycle before 2007 or one that is older than 2007, you will need to pay emission zone charges in many cities. In fact, they are likely to be banned from cities completely. How do we deal with this in relation to sustainability and the future?
2. The Whitstable and Herne Bay group of charity motorcycle riders, which the author rides with, meet at Christmas to take toys to the local children's hospital. About 200 raucous motorcycles do the 10-mile journey. People come of their doors to wave and make contributions to the collection fund. The children stand at the windows of the hospital as they hear the approach of the motorcyclists dressed in Santa costumes. How is this going to sustain into the future when all motorcycles are electric?

Tech Note

A Boeing 747 jet passenger airplane has a fuel capacity of 238,604 liters. It has a flight range between refills of 7790 miles. A return trip to Australia of 21,000 miles will therefore require three tanks full.

INSURANCE

If you are working as a motorcycle engineer, you are advised to take out professional indemnity insurance along with public liability and employer liability coverage. This should cover you for claims from neighbors and accidentally damaging the environment too.

Chapter 20

Brief history of motorcycles

Motorcycles are part of our transport history, going back to the time when roads were intended for horse-drawn vehicles and those traveling on foot and horseback. In the UK we have had a road system since before the Romans invaded in the early first century. In fact, there was a worldwide road system catering for traders in spices and silks – some adventure motorcyclists travel these routes for the feeling of oneness with the past. The Romans created a system across the UK, and other countries, to enable the transportation of metals, other minerals, and slaves back to Rome.

The roads at the time of the invention of motorcycles were simply tracks of dirt, stones, and gravel. This was not conducive to two-wheel travel at all; hence, the stable four-wheel car quickly took over for the most part.

> **Tech Note**
>
> Before designing something, it's a good idea to look at what has gone before. No point in reinventing the wheel, but you can probably make it lighter or rounder.

However, the motorcycle offered then, as it does today, a number of advantages over the car. It is cheaper to make, having fewer parts, though the current prices do not always give this impression. Motorcycles are easier to park or store, as they take less space; they can get through smaller gaps in traffic, but towns that have introduced "traffic calming" prevent this. Still, motorcycles always have been, and always will be, a delightful form of relatively low-cost personal transport. Let's have a more detailed look.

The first motorcycle is credited to have been made by Gottlieb Daimler and Wilhelm Maybach in 1885. This is known as the Einspur. Interestingly, the locations of Baron von Drais and Gottlieb Daimler and Nicolaus Otto are very close together in what is now Germany. Einspur could be considered a Draisienne bicycle with an Otto engine. After producing one. Daimler quickly moved onto making cars – what we now know as Mercedes Benz.

This first motorcycle had a wooden wheel, steel cart tires, and an engine mounted where the rider would expect to sit. Although it was impractical, it set the basic layout for what could be done.

Like all things, the devil is in the detail – in this case, the details of the construction materials. At the same time hollow steel tube, pneumatic tires on spoke wheels, and roller chains were being developed. So, we had a whole rush of motorcycles that could actually be ridden. Technically, Einspur was capable of 8 mph, as they would say in the current motorcycling magazines – manufacturer's figures. Nine years on, in 1894 Hildebrand and Wolfmuller (H&W) produced a 1500-cc motorcycle with a steel frame and pneumatic tires that did 25 mph – now we are motorcycling. The problem was braking and handling. The H&W had a high mounted engine, giving a high center of gravity (CoG) and no brakes to speak of.

The Werner Bros. of Paris then entered the stage. They mounted the engine low down in the frame and added rim brakes. With lightweight, smaller engines and belt drives, these were easy to use. Versions called motorcyclettes, what we would call mopeds, were made with pedals and a chain drive. These would have been really good in the very busy streets of Paris as the timeline moved into 1900 with lots of motor cars, as well as horse-drawn carriages and bicycles, and the new railways. Paris was, as it still is, the home of the modern and the stylish, the Eiffel Tower having been built in 1887 and attracting visitors from all over the world. To put this into context, Paris had traffic lights in 1912; it was 1926 before they were used in London. Horses continued to be the most popular form of transport in London until the mid-1920s, unlike Paris that had cars, taxis, and motorcycles by the thousands.

Back to 1900. With the technology and engineering of motorcycles now firmly established, lots of engineers, both professional and amateur, started to throw their cap into the ring with an intention of satisfying a motorcycling need and making money. It's worth mentioning that at this time there were a lot of companies with words like manufacturing and machining in their names, mostly in the Midlands of England and the Midwest of the United States, as well as Paris and the area around Stuttgart in Southern Germany. These companies would make anything, for example: pots and pans, sewing machines, bicycles, agricultural equipment, and garden gates. Motorcycles offered the main chance; riches could be made.

Motorcycles, if they are to work well, need to be made with a high level of precision. The engineering firms that had produced guns knew about mass production to accurate tolerances. Firms such as Birmingham Small Arms (BSA) were in a great position. They made bicycles, motorcycles, and cars until the parts were separated and they were taken over by bigger firms. BSA started in 1861 and eventually closed in 1973. Royal Enfield Motorcycles, also manufacturers of guns, was founded in 1901; they are still in business but only with motorcycles. FN which stands for Fabrique

Brief history of motorcycles 333

Figure 20.1 Draisienne, also called a dandy horse. The rider sat on it and paddled it along with his legs. The start of the two-wheel revolution.

Figure 20.2 Boneshaker bicycle in use on the Tweed Run in London. This is an event for dapper bicycle riders. Now there's pedals – feet off the ground.

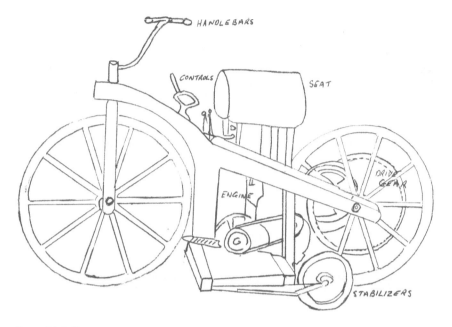

Figure 20.3 Einspur – the first motorcycle – produced by Gottlieb Daimler in 1885. Notice the positions of the components – not too unlike current designs.

Nationale de Herstall, was a gun maker from 1889 to the present day and made motorcycles up until 1967. The history of the Royal Enfield Company, and it has had a number of different names, dates back to 1851. They originally made sewing machine needles, then bicycles, then parts for the British government's munitions factory at Enfield in Middlesex.

PURPOSE AND USE

Why do we have motorcycles is part of the question of the history of motorcycles. Motorcycles have developed in a number of different ways and styles. And currently the trend is to retro-styling, looking backwards to look forwards. Motorcycles have the advantage of a high power to weight ratio but the problem of road adhesion and stability because having only two wheels cancels out a lot of the advantages. Scientists and mathematicians have spent years trying to work it out, but there is no clear answer.

Motorcyclists like speed, so if we look at land speed records you would expect motorcycles to have the record, but no. Currently the speed record for a motorcycle is 376 mph; for a car it's 763 mph, over double that speed. You can have all sorts of arguments as to why this is not fair; so, go back to 1903, the motorcycle record is 64 mph, the car 85 mph. What is perhaps more surprising is that the railway speed record – a train – was 77 mph in this same year. The railway lines were being built alongside the roads, so there must have been cases of motorcyclists being overtaken by a train with his or her mates jeering out of the window. This led to a number of unofficial records of motorcyclists trying to beat trains between major cities. In 1903 the speed limit was 20 mph; there was no speed limit for trains.

Since the very beginning motorcycles have had a split personality. They have been cheap everyday utility transport for individuals and at the same time very expensive Sunday morning toys for rich people. They have been raced on circuits and used to explore mountain ranges.

WORLD WAR I

Suddenly the motorcycle had a purpose. When World War I broke out in 1914 troops called the British Expeditionary Force (BEF) were sent to France: 247,000 men, 827 cars, and at the end of the list 15 motorcycles. Obviously, the motorcycles were either a late addition or were designed to satisfy some sort of special request. The motorcyclists of those 15 machines were involved in the first fight with the Germans – they won. Suddenly the War Office decided that they needed more motorcyclists. They put out a call for volunteers to sign up with their own motorcycles. Thousands joined and became part of the Royal Engineers. The author's maternal grandfather was involved with this initiative and I'm very proud of him.

One of the main motorcycles makers was Triumph, who supplied 30,000 Model Hs. There were several thousand Clyno motorcycle combinations made in conjunction with Vickers. These usually had machine guns fitted to the sidecar, even though some of the Australian solo riders could shoot while riding at speed (sounds like something out of *Crocodile Dundee*). Douglas was also contracted to supply what was initially going to be a small number of 300 motorcycles per month – the total eventually came to 70,000. Royal Enfield – just called Enfield at that time – also supplied many motorcycles and sidecars with machine guns attached. Another big supplier of motorcycles was Scott; they turned all their production to the war effort. Interestingly the Scott motorcycle had to be designed and marketed as a luxury wheeled horse for the Edwardian gentleman – what we would now call a Sunday morning motorcycle. The dispatch rider volunteers were allowed to ride their own motorcycles; in fact they had been called to volunteer this in way. Many of these were riders of Scotts.

The main use of the motorcycles was in carrying messages – difficult to think, no telephones, and radios took three men to lift them. They had messages written on paper. They also took carrier pigeons in baskets to the front line. The army used thousands of pigeons. Each day the motorcyclists would ride into the war zone with the precious cargo. I'm sure that lots of current motorcyclists would do this now if needed. When the Americans joined the war in Europe in 1917, they brought Harley Davidsons and Indians – big-engined V-twins similar to how they are now.

It's worth noting that World War I was the biggest spending that the government had had. The government had factories producing munitions for the army but had never before needed such a large volume of items – so large that they had to go to independent manufacturers to supply the goods. The costing for these motorcycles and all the other engineered items was done on a cost-plus basis. That is, the government paid for the item and gave an extra 5% for profit. This helped keep the engineering factories going during the war maintaining jobs – mainly for women, who were not fighting.

Figure 20.4 Vintage Velocette.

Figure 20.5 Cockpit of vintage motorcycle: left-hand side is clutch – obscured and advance and retard lever; center is brass knob for steering damper and bulb horn; right-hand side is choke lever and just visible end of front brake lever. On top of petrol tank is quick-lift, nondetachable filler cap.

Figure 20.6 Vintage carburetor.

Figure 20.7 Vintage dynamo.

Figure 20.8 Vintage single-sided front drum brake.

BETWEEN THE WARS

As the world started to return to peacetime, now the 1920s, there were hundreds of small companies setting up to produce cars, bicycles, and motorcycles. In the UK, the number was about 2000. World War I had changed the social system and brought a lot of new ideas and technology into common usage. The developments were enormous, bringing changes to motorcycles and how they were used. Let's look at some of the factors:

- Five million men and women who had fought in World War I were being demobilised and seeking work.
- Many of those who had previously been the upper classes now had lost wealth and property.
- The Royal Engineers and other regiments had given engineering training to tens of thousands of men and women.
- There was a desire by the people who had fought in the war for the world to be a better place.
- There was a high level of inflation – about 15% – in the UK, but in Germany there was hyperinflation.
- The companies that had built motorcycles and other munitions were in a strong position to move forward.

In other words, it was all systems go for the motorcycle industry. As the years progressed the UK and other countries moved into a recession until 1939 when World War II broke out. Oddly, this appears to have had a fairly positive effect on the motorcycle industry in terms of the generation of new models.

The need for speed was indicated by the land speed record being broken 22 times between the two wars. There was a profusion of what we now call hyperbikes and high-class gentleman's wheeled-horses, as well as very cheap commuter machines. The Brough Superior with its 1000-cc JAP engine, was top of the shop alongside a range of Vincents. These hyperbikes could cruise at 100 mph when most others were struggling to do 50 mph. The 1937 model of the Brough Superior broke the motorcycle speed record of 169 mph.

The middle-priced motorcycles – Norton, Triumph, AJS, Matchless, and Royal Enfield – all started to chase the 100 mph figure. In 1937 Triumph produced the 500-cc Speed Twin, the same speed as the Brough Superior but with half of the capacity and at a fraction of the price. A modified version of the original speed twin engine is still in use today in the current models with that name.

Also available worldwide were motorcycles from manufacturers in America and Continental Europe, marques like BMW, CZ, Indian, Harley Davidson, Jawa, Ducati, Puch, Peugeot, New Map, Zundapp, DKW, NSU, and many more.

WORLD WAR II

At the start of World War II, the UK was rather better organized, and the War Office knew the benefit of motorcycles. Again, there was a call for motorcycle enthusiasts with mechanical knowledge – the author's father was one of the thousands who responded, quickly becoming an instructor at Aldershot. The riders were taught to ride by numbers: 1. Stand by the motorcycles. 2. Switch on and kickstart. 3. Hold in clutch and select gear, etc. The instructor was given a faster machine to be able to lead the group. Harley Davidsons with footboards were the machines of choice for instructors.

Most of the British makes of motorcycles were used, and special lightweight folding motorcycles were used to be parachuted into war zones. This was another financial shot in the arm for the motorcycle industry, and after the war many of these machines were available on the used machine market. Between the start and end of World War II prices had inflated about 300%, so ex–War Department machines were highly desirable for the commuter market in the late 1940s and early 1950s.

THE GOLDEN ERA

The 1940s, 1950s, and 1960s was the golden era of motorcycles. Motorcycles were readily available, reasonably reliable, and at affordable prices with good roads to ride them on. The supply of new cars was limited by the rationing of steel until 1957.

World War II led to many developments both in motorcycle technology and manufacturing processes. Let's list a few:

- Rear suspension now included shock absorbers instead of the solid rear end.
- Cush drive hubs were preventing chain snatch.
- Telescopic front forks were replacing the girder ones.
- There was a choice of coil ignition and magnetos.
- As higher-octane petrol became available compression, ratios were increased to provide more engine power.
- All -aluminum engines were introduced.
- Unit construction of engine and gearbox in a shared casting led to less weight and a stiffer assembly.
- Unit construction also allowed the use of smaller frame tubes.
- Tires were designed for their road holding capabilities.
- Automatic and semi-automatic lathes, called capstan lathes, were used in many stages of production.
- Broaching and form grinding could be used on gears.
- Heavy-duty metal pressing was possible.
- Painting and finishing processes had become automated.

These were major shifts in design and manufacturing from the pre–World War II period, and the driving force was production engineering.

The period 1946 to 1964 was also known as the baby boomer period because of the great increase in the birth rate. The year 1948 saw the start of the National Health Service (NHS).

Significantly, the music changed. Bebop, or simply bop, a style of jazz, came in the 1940s, as did the big band sound. This was a very upbeat set of sounds, with varieties including swing and pop. This transformed into rock and roll with Elvis Presley, the main protagonist. The motorcyclist tended to go for the rock and roll; the scooter riders the Mod music – a more foot-tapping regular beat. Mod music leaders were The Lambrettas, Small Faces, and The Jam.

The production methods led to mergers of manufacturers, but brand names were kept – referred to as marques. The same names from pre–World War II by and large remained in the 500-cc/650-cc class, viz:

Figure 20.9 Six-cylinder Honda.

Figure 20.10 Six-cylinder exhaust pipes.

Norton, Triumph, BSA, Matchless, AJS, Royal Enfield, and in the smaller engined class Francis Barnet, James, and a number of other two-stroke engined machines. The main scooters were, as now, Lambretta and Vespa, imported from Italy.

In 1962 Triumph took the motorcycle speed record to 224 mph at Bonneville – hence the famous name *Triumph Bonneville* which still continues now.

1969 – THE SEA CHANGE

It has to be said that motorcyclists were, and maybe still are, traditional in their thinking, and there is nothing wrong with that, I hasten to add.

It should also be remembered that in the early postwar years many men were still feeling the scars of the Japanese prisoner of war camps. In short, the cheap imports of small motorcycles from around the world were met with disdain, to put it politely. But in 1969 that all changed. The Honda CB750 four-cylinder motorcycle changed the color of the sea forever. It looked like any other what we would now call naked, or classic, motorcycle, but it was nothing like them in the best of all ways. Again, the devil was in the detail.

Let's discuss Mr. Honda for a moment. He was the son of a blacksmith on the side of Mount Fuji in Japan. He served a five-year apprenticeship as a motor mechanic. During World War II he was making piston rings for Toyota. At the end of World War II, he set up a business making electric-powered bicycles, eventually moving on to motorcycles and cars. His philosophy was one of continuous improvement. The Japanese engineering philosophy is about total quality management (TQM). The Japanese monetary philosophy is about long-term gains and a small percentage for many years – the opposite of the western banking culture.

What the CB750 offered was:

- Four-cylinder OHC – compared to the twin cylinders of many motorcycles, it had twice as many firing strokes
- Horizontally split unit construction prevents oil leaks
- Front disc brake
- Electric start

Traditional British bikes, like my BSA, emptied their gearbox oil on the floor like a badly trained puppy every time it was left – I carried a can of oil to top up, even though all the gaskets and seals had all been replaced. The electric start meant no more bruised legs as the kickstart sprang back. And wow, a real 120 mph, not just downhill with the wind behind you.

There was in fact no new technology. We've had four-cylinder motorcycles since 1900, horizontally split crankcases were to be seen on most cars, disc brakes were invented by Lanchester in 1912 and used on high-performance cars, and the electric start – or dyno-start – had been used on cars in the 1920s.

The Honda CB750 was in fact very traditional in looks, but it was really well made. The detail was all correct and it worked. Mr. Honda had perfected piston rings, so he perfected all the bits to make the motorcycle work. It is also interesting to note that at this time Mr. Shimano was perfecting the bicycle free-wheel – concentrating on the one component, then branching out into all the other moving parts after the one part had proved its value – or should I say justified its price point (none of these parts are cheap).

In the 1970s and 1980s motorcycling was not popular. This is probably because cars had become cheaper and the legal need for a 30% deposit to buy one on hire purchase had been removed. The end of the 1980s saw a financial recession which lasted into the 1990s.

This period also highlighted the differential pricing of motorcycles between the UK and the rest of Europe. Motorcycles could be bought 30% cheaper in France and Germany than in the UK. It became quite normal to drive across the Channel to save on a new machine. You could also save 30% or more on wine, beer, and cigarettes on the same trip.

As the timeline enters the 1990s we have motorcycles that work – as they say, they do what it says on the box. These machines worked reliably with minimal maintenance and have great performance and economy. The designers started looking at aerodynamics and weight. We now saw:

- Full fairings for aerodynamics and rider weather protection
- Lightweight alloy frames
- Lightweight alloy wheels
- Disc brakes, front and rear
- Aerodynamic fitted luggage systems
- Single-sided rear trailing arm suspension with monoshock absorbers
- Upside-down front forks

The year 1990 saw the birth of the Kawasaki ZZR 1100, the hyperbike that would hold the world's fastest production motorcycle title for the next five years.

THE MILLENNIUM

The generation referred to as the millennials were born between 1980 and 1984. In 2000 they were the young motorcyclists between 16 and 20 years old. For this generation, just about any form or style of motorcycle was available. In 1999 Suzuki introduced the almost 200-mph Hayabusa GSXR1300. This was in answer to Kawasaki's ZZR 1100 and ZX1200R and Honda's record-breaking Super Blackbird 1100. Hayabusa is a kind of hawk that eats blackbirds – a good marketing play on words. In 2001 the motorcycle industry voluntarily agreed to limit the maximum speed to 186 mph. In 2020 just about all motorcycle manufacturers produce a machine in each of the market segments, from hyperbike to scrambler. The speed limiters can be switched off with a laptop and appropriate software. The fastest motorcycle currently is the Kawasaki Ninja H2. It will do 250 mph standard – faster than the streamlined Triumph Bonneville and the £3M Bugatti Veyron.

Figure 20.11 Margate Meltdown – part of lineup of a wide range of motorcycles.

TIMELINE

This timeline is a guide to enable the reader to see the developments of the motorcycle along with other developments in engineering, science, and technology and in relationship with other major sociopolitical events. History is mostly unclear – even if we were there on the day, we still may not have seen the actual things that triggered an invention or other change; however, viewing it in context helps to define the moment and may help to provide an insight for future motorcycle engineers. All dates are approximate – the timeline reflects commonly accepted data.

Figure 20.12 Roads closed for Margate Meltdown – motorcycles and scooters ride out from Ace Café in London to Margate in Kent; this is a great day out for any two-wheel enthusiast.

Brief history of motorcycles 347

Time-Line

Date	Motorcycle development	Science, Technology and Arts	Other socio-political events
1600 – 1800	Goods and passenger carrying coaches are developed to a high level with regular services between major towns. Springs, bearings and metal tyres are in use. These horse-drawn vehicles have efficient brakes and steering systems. So, the basic materials and manufacturing capability for new forms of transport now exists. What doesn't exist is the understanding of balance for bicycles until this is found with the *Hobbyhorse* in the safe environment of a public park, although one can imagine boys playing *hoop and stick* with old cart metal tyres.	The start of the new century – 1600 marks the beginning of the scientific revolution – Galileo and Bacon start to examine nature; they use *scientific method* as we use today 1769 Nicholas Cugnot builds steam tractor; in 1771 he crashes it -the world's first motoring accident 1744 James Watt patents a steam carriage; but does not make it	Re-organisation of postal services, Enforced Sunday Observance, First Turnpike Act – charging road user, Bubonic Plague, Fire of London, Lloyds Insurance started, Bank of England founded. 1760-1830 Industrial Revolution 1776 American Independence 1789-1799 French Revolution
1818	Draisienne, or *Hobbyhorse* bicycle of Baron von Drais de Sauerbrun. This forms the basis for bicycles and motorcycles	1801 Dick Trevithick drives his steam carriage up Cambourne Beacon in Cornwall. In 1803 he builds the London Steam carriage	1830-1890 Universities start to open up higher education for all
1839	Kirkpatrick Macmillan make rear-driven bicycle		1838-1848 Chartism, votes for the working class
1863	Velocipede *Boneshaker* bicycle by Pierre and Ernest Michaux		
1860 – 1880	A variety of cycles are built by various people, these include: monocycle, dicycle, tricycle and quadricycle	1823 Introduction of Rugby football	1861-1865 American Civil War
1864	Siegfried Marcos – Austria invents the petrol engine	1845 R W Thomson invents the pneumatic tyre	
1869	Michaux-Perreaux a firm of bicycle makers in Paris produce a steam powered motorcycle. Steam power was now to be commonly found in trains and ships and some road coaches	1847 Institution of Mechanical Engineers is founded – to share Mechanical Engineering Knowledge	In the 1860s: The Turnpike Trusts started to put up their prices. The National Tolls were 15p for horse drawn vehicles and £2 for steam vehicles. With £5 fines for non-observance – this was equal to a year's wage for a working person.
1870	The Ordinary Bicycle or *Penny-farthing* of J K Starley and W Hillman is made with tension-wire spokes	1850 Introduction of Golf to England	
1876	Otto makes 4-stroke petrol engine, these are later made in Manchester by Crossley Brothers		

(*Continued*)

Date	Motorcycle development	Science, Technology and Arts	Other socio-political events
1877	J C Garood introduces tubular frame construction for bicycles		1865 The Locomotive Act, referred to as the Red Flag Act effectively stops the development of all motor vehicles. The law required: A person must walk in front of the vehicle carrying a red flag, Two people must be in the vehicle, Lights must be fitted, Speed limit of 2mph in town, 4mph in the country, The vehicles were subjected to tax, A horse rider could request them to stop
1877	Ball and roller type bearings are introduced		
1877	J K Starley makes the Coventry lever tricycle		
1877	J K Starley makes The Royal *Salvo* tricycle with differential gears		
1879	H J Lawson makes the *Bicyclette* with rear chain drive		
1880	Hans Renold makes the bush roller chain for J K Starley		
1881	K White and G Davies make the free-wheel mechanism		
1885	J K Starley makes the diamond shaped frame for the Rover Safety bicycle – what we know as basis for the modern bicycle	1886 First car built	
	The same year Gottlieb Daimler produces a petrol engine motorcycle	1887 First production car sold by Benz	
1888	J B Dunlop makes the pneumatic tyre		1870s The roads are started to be 'dis-turnpiked' as most traffic had moved to the railways which were being built alongside the roads
1894	Hildebrand and Wolfmuller produce 1500cc motorcycle	1889 Daimler make first production car	
1897	A M Reynolds and J T Hewitt make butted steel tubes		1878 Red Flag Act amended
1899			
1900	Raleigh introduce the all-steel safety bicycle. The cycle motor is introduced		1889 - 1902 Boer War
1901	Werner designs a motorcycle with the engine fitted low between the wheels		
1902	Indian produce a single cylinder motorcycle		1903 Speed limit raised to 20mph
1903	Motorcycle speed record of 64mph by Glenn Curtiss on a Curtis V-2 of 1,000cc		
1905	FN – the Belgium gun makers – make a 4-cylinder motorcycle		
1906	Lanchester invents disc brakes and the first roadside petrol pump is set-up		
1907	Supercharger first used		

Brief history of motorcycles 349

Date	Motorcycle development	Science, Technology and Arts	Other socio-political events
1909	Indian and Pierce (Arrow) make a 4-cylinder motorcycle		
1910	Scott produce their water cooled 2-stoke which is used by the army in both WW1 and WW2 – it runs almost silently - stealth		1914-1918 World War 1
			1912 First traffic lights are installed in Paris
1911	The Bowden electric lighting generator is produced		
1914	Peugeot make vertical twin with 4-valve cylinder head		1917 Russian Revolution
1922	It is recorded that there are over 2,000 individual manufacturers of motorcycles and cars in the UK	1922 BBC formed	
1930s	At this time motorcycles had evolved into the shape and style that we recognise today; indeed, many manufacturers are copying shapes from this era into modern-retro. The hyperbikes of that time, the 1,000cc Brough Superior and Vincent were capable of 100mph, albeit at a price. The JAP engined Brough Superior held the speed record of 169mph in 1937. Then in 1937 Triumph introduced the 500cc Speed Twin making this kind of performance available, with a price ticket that was affordable, to a large percentage of motorcyclists with half of the capacity and less weight. At the same time the middle price ground also had Norton, Royal Enfield, Matchless, AJS and a myriad of lower powered models such as James and Francis Barnett.		1929 till late 1930s The Great Depression

1930s The A56 from Burnley to Manchester and Liverpool was the last road to be metalled - Tarmac |
| 1939 – 1945 | Most motorcycle manufacturers switch production to war effort of both motorcycles and munitions | 1945 Fist Atom Bomb used
1948 National Health Service
1955 Independent Television formed | 1939-1944 World War 2 |
| 1940s, 1950s and 1960s | The Golden Era of motorcycling epitomised by the Norton, BSA and Triumph models that still set the hearts of young motorcyclists racing. In 1962 Triumph took the speed record to 224mph at Bonneville – hence the famous name *Triumph Bonneville* name which still continues | 1957 Rock and Roll – Elvis Presley
1962 Beatles pop group
1969 Moon Landing
1976 Concorde supersonic aircraft
1978 Water Speed Record, 317mph by Ken Warmby | 1947-1991 Cold War
1955-1975 Vietnam War
1963 First reported traffic jam – 35 miles into Torquay, Devon |

(*Continued*)

Date	Motorcycle development	Science, Technology and Arts	Other socio-political events
Late 1960s – 1970s	The motorcycle changes to meet the modern world with the introduction of Japanese models from Honda, Kawasaki and Suzuki. Electric start, horizontal split – non-leaking crankcases, flashing indicators, low emissions and very light weight		1982 Falklands War
1986 M25 London Orbital Motorway is opened
1988 Longest reported traffic jam of 117 miles on M4 from London to Bristol
2010	Again, at Bonneville Salt Flats in Utah, USA the speed record, which still stands, was taken to 376mph with a twin Suzuki engined streamliner.	1997 Land Speed Record, 763mph by Andy Green in Thrust SSC	
2010 - 2020	A re-birth of motorcycle sales, an increase in motorcycle licence holders – mainly over 50years old males. The rise of the Adventure model motorcycles	2019 Air Speed Record, 2,193mph in Lockheed Blackbird	2020 Brexit

Apprenticeship standard for a motorcycle technician

DETAILS OF STANDARD

Overview of the role

A motorcycle technician services and repairs motorcycles, motorized scooters, all-terrain vehicles (ATVs), tricycles, and vehicles where the rider sits astride the frame, within either the franchised or independent motorcycle sector.

A motorcycle technician works on all the systems of the motorcycle. The nature of the work ranges from replacing components through to rectifying complex faults with the aid of specific diagnostic methods and equipment. Motorcycles require individual setup and adjustment in relation to rider requirements, which include leisure, commuting, commercial, and competition use.

Today's motorcycle technician has to demonstrate expertise in the technical side of their role. They need strong problem-solving skills and a good grasp of the theoretical, practical, and safety aspects of motorcycle systems.

They must be able to work independently and operate as an effective team member, understand how their workshop and a motorcycle business functions from a commercial perspective, develop good customer handling skills, and identify ways in which they can work efficiently.

Level: Level 3
Duration: Typically, 3 years

Entry requirements

Individual employers will set their own selection criteria for the applicants. However, it is recommended that to optimize the chance of selection, a candidate can demonstrate an interest in the motorcycle environment. It is also recommended that the employer assesses the candidate's ability to demonstrate mechanical skills and communicate both orally and in writing.

Qualifications

Apprentices without level 2 English and maths will need to achieve this level prior to completion of their apprenticeship.

Knowledge

Technicians must have knowledge and understanding of the types, application, and unique characteristics of motorcycles, scooters, tricycles, ATVs, and quad all-terrain vehicles (EU Type Approval L category) where the rider sits astride the frame and of motorcycle technologies:

- Frame construction and knowledge of welding and brazing methods
- Handlebar direct control, steering setup, and geometry systems
- Differing front and rear suspension systems, including setup and adjustment
- Front and rear and combined braking systems: disc, drum, cable, and hydraulic
- Wheels and tire specifications, characteristics, and uses to include competition, off-road, road use, and touring
- Engine/power source, including two stroke, four stroke, single and multiple cylinder construction, and electric propulsion
- Cooling and lubrication systems, including air and liquid cooled and wet and dry sump engines.
- Fuel and ignition systems, including carburetor and injection plus ECU
- Intake and exhaust systems, including emission
- Transmission to include chain, belt and shaft drive, and CVT drive-line systems
- Electrical systems, including engine management, lighting, monitoring and instrumentation, and security and accessory fitting (including electronic fault diagnosis)
- How to service, inspect, and maintain motorcycles to ensure safe operation meeting all legal, licensing, and customer expectations and requirements
- Understand the benefits of on-road testing to diagnose faults and to verify correct rectifications
- Diagnostic principles, troubleshooting, logical problem-solving, and repair techniques
- Health and safety knowledge and environmental awareness to carry out work safely
- Emerging technologies and legislation, including electric motorcycle developments and the impact they will have on the knowledge and skills motorcycle technicians will require in the future

Skills

- Perform the fundamental engineering tasks that most procedures include, such as cutting, drilling, filing, removing and replacing bolts, screws and clips, replacing seals, extracting damaged fasteners, and using fabrication skills
- Assist in upholding high standards of safety and efficiency in the workshop, adhere to the requisite business processes (e.g., environmental awareness, health and safety practices, record keeping, and customer contact), and standard workshop
- Safely secure motorcycles to ramps and use specialized supporting stands in order to remove road wheels and major components, including brake systems, suspension, and drive (chain, belt, or drive shaft)
- Successfully inspect and prepare a motorcycle to the required quality standard for handover to the customer (e.g., following a service, complex repair, pre-delivery inspection by checking work against)
- Service and maintain motorcycles in line with manufacturer specifications logged in manuals and online
- Remove, repair, and replace components in line with manufacturers' defined instruction
- Use diagnostic methods and mechanical and electrical measuring tools and equipment to check compliance and rectify faults
- Investigate symptoms of motorcycle fault(s) and identify the underlying causes prior to repair
- Access specific information (e.g., motorcycle repair information, wiring diagrams, maintenance tables, technical production information, safety recalls)
- Apply advanced diagnostic principles, logical problem-solving techniques, and complex

Behaviors

- Work cohesively with team colleagues and company ethics to ensure quality workmanship
- Ensure all work processes are carried out safely and report any concerns or
- Communicate effectively and treat customers with respect when discussing topics that will support the process of diagnosing and rectifying faults and specific component setup requirements
- Behave in accordance with the values of the company they work for, operate as an effective team member, and be able to manage own time effectively
- Take responsibility, being honest and accountable for own actions and work

Progression

On completion of this standard individuals will be able to work towards higher levels of technical training, including master technician or professional accreditation, along with supervisory and managerial positions. In addition, manufacturers may offer their own specific product-related advanced training that will be made available to individuals working in the franchised industry or as continued professional development (CPD). On successful completion of this apprentice standard, the individual will be eligible to apply to join the Institute of the Motor Industry (IMI) Professional Register, a register of qualified time-serving individuals in all aspects of the motor industry, enabling the public and employers to verify the person has qualified as an apprentice.

Glossary

This section defines a number of the words and phrases used in motorcycle engineering and general motorcycle usage, including some of the specialist racer and enthusiast vocabulary and jargon.

12-hour/24-hour: a race lasting 12 or 24 hours; the winner is the one covering the greatest distance
Acceleration: rate of increase of velocity
Accessories: anything added that is not standard on a motorcycle
Add-ons: something added after a motorcycle is made
Adhesion: how the motorcycle holds the road
Alignment: position of one item against another
Alloy: mixture of two or more materials; may refer to aluminum alloy, of an alloy of steel and another metal such as chromium
Ally slang for aluminum alloy
Atom: single particle of an element
Barrel: a hollow cylinder, also called a pot
BDC: bottom dead center
Bench: working surface; also flow bench and test bench
Beta version: test version of software or product
Bore: internal diameter of cylinder barrel
Brooklands: first purpose-built racing circuit at Weybridge in Surrey with banking and bridge; Monza in Italy was copied from this
Carbon fiber: like glass fiber but uses very strong carbon-based material
Chicane: sharp pair of bends – often in the middle of a straight
Chocks: tapered block put on each side of a wheel to stop the motorcycle from rolling
Circuit: race circuit
Clerk of course: most senior officer at a racing event – person whose decision is final, although there may be a later appeal to the various governing bodies
Code reader: reads fault codes in the ECU of the particular system

Composite: material made in two or more layers – usually refers to carbon fiber, but may include a honeycomb layer
Condensation: changes from gas to liquid
Contraction: decreases in size
Corrosion: there are many different types of corrosion, with oxidation or rusting being the most obvious
Cushion: section of seat to sit on, also cush-drive in the rear wheel
Dashboard: instrument panel
Cockpit area around handlebars and speedo
Density: relative density; also called specific gravity
Diagnostic: equipment connected to the system to find faults
Dive: motorcycle goes down at front under heavy braking, or rider is thrown off
Drag strip: flat and smooth section of race track – fast section of road
Epoxy: resin material use with glass fiber materials
"Esses": one bend followed by another
Evaporation: changes from liquid to gas
Event organizer: person who organizes the race or other event
Expansion: increases in size
Fast back: extended rear mudguard
Flag marshal: marshal with a flag
Flag: checkered flag, black flag, red flag, and other colors used for different purposes
Foam: material used for making seats and other items
Force: mass multiplied by acceleration
Friction: resistance of one material to slide over another
Frontal area: (projected) area of front of motorcycle and rider
Gelcoat: a resin applied when glass fiber parts are being made – it gives the smooth shiny finish
Glass fiber: lightweight mixture of glass material and resin to make vehicle body
Heat: a form of energy; hotness
Hill climb: individually timed event climbing a hill
Inboard: something mounted on the inside of the drive shafts such as inboard brakes, usually lowers unsprung weight
Inertia: resistance to change of state of motion – see Newton's Laws: inertia of motion and inertia of rest
Kevlar: fiber super-strong material, often used as a composite with carbon
Machine the motorcycle
Marshal: person who helps to control an event
Mass: molecular size, for most purposes the same as weight
Metal fatigue: metal is worn out
Molecule: smallest particle of a material
Monza: World's second banked race track in Italy; copy of Brooklands

Newton's Laws: First Law – A body continues to maintain its state of rest or of uniform motion unless acted upon by an external unbalanced force

Second Law – The force on an object is equal to the mass of the object multiplied by its acceleration (F = Ma)

Third law – For every action there is an equal and opposite reaction

Nose cone: detachable front body section of streamlined motorcycle or sidecar outfit – may include a foam filler for impact protection

"O" rings: rubber sealing rings

Off-roader motorcycle for going off-road; or off-road event

Original finish: original paint work, usually with reference to historic or vintage motorcycles

Outboard: something mounted on the outside of the drive shafts such as brakes, usually increases unsprung weight

Oxidation: material attacked by oxygen from the atmosphere such as aluminum turns into a white powdery finish

Paddock: where teams and motorcycles are based when not racing

Parent metal: main metal in an item

Pot: another name for cylinder

Power: work done per unit time: HP, BHP, CV, PS, kW

Prepping: preparing the motorcycle for an event

Prototype: first one made before full production

Start ramp downward incline used at the start of some events

Regs: racing regulations

Ride height: height of motorcycle seat off the road; usually measured from road to the top of the saddle, also the clearance from the road

Rings: piston rings

Rust: oxidation of iron or steel – goes to reddish color

Scrutineer: person who checks that a motorcycle compiles with the racing regulations, usually when scrutineered the motorcycle has a tag or sticker attached

Skid: motorcycle goes sideways – without road wheels turning

Speed event: any event where motorcycles run individually against the clock

Spine: backbone-like structure

Sprinting: individually timed event starting from rest over a fixed distance

Sprung weight: weight below suspension spring

Squat: motorcycle or rider goes down at the back under heavy acceleration

Squeal: high-pitched noise – usually from brakes

Stage event: when the event is broken into a number of individually timed stages, the motorcycles start the stage at preset intervals (typically 2 minutes)

Stall: involuntary stopping of motorcycle and rider – usually due to engine cutting out – in trial events this attracts points

Steward: a senior officer in the organization of the motorcycle event

Straw bales: straw bales on side of the track for a soft cushion in case of an off
Stress: force divided by cross-sectional area
Stripping: pulling apart
Stroke: distance piston moved between TDC and BDC
Swage: raised section of metal panel
Swage line: raised design line on metal panel
TDC: top dead center
Temperature: degree of hotness or coldness of a body
Test bench: test equipment mounted on a base unit
Test hill: a hill of which the gradient increases as the top approaches; originally the test was who got the farthest up the hill; Brooklands Test Hill is occasionally used
Torque: turning moment about a point (Torque = Force × Radius)
Transporter: vehicle to transport motorcycles and team equipment to events
Team bus: vehicle used to transport the team of riders and officials to events
Tire wall: wall on side of track built from tires – giving a soft cushion in case of an off
Unsprung weight: weight below suspension spring
Velocity: vector quality of change of position; for most purposes the same as speed

Index

ABS 192
Air filter 51
Air:petrol ratio 50
Alloying metals 115
Alternator 201–202
Aluminium alloy 119–120
Aluminium brazing 149
Amps 255–256
Annealing 113
Apprenticeship standard for motorcycle technician 353–356

Balance sheet 246–247
Balancing 30
Battery 36, 197–201
BHP per ton 5
Block assembly 29
Bomb calorimeter 7
Brake fluid 190
Brake lines 190
Braking system 176
Brazing 146–147
Bronze welding 142

Cables and connectors 206
Capacity 19
Capacity and volume 253
Casting 112
Castor 161
Chassis earth 205
Clutch 259–263
Coil ignition 36
Cold rolling 112
Columbus tubing 120–121
Combustion 25
Composite materials 225

Compression ratio 20, 22
Connecting rod 10
Coolant 66
Cooling 55–74
COSHH 86
Crankcase 8
Crankshaft 10
Cruciform frame 123
Customising 275–298
Cylinder block 8
Cylinder head 10, 27

Data 239–247
Diamond frame 123
Diesel 18
Digital thermometer 170
Distributor 41
Double flare 190
Durometer 171
Dwell angle 43

Einspur 334
Electric motorcycles 113
Electrical hazards 85
Electronic ignition 43
Electronic tyre pressure gauge 170
Energy from fuel 6
Engine 1
Engine performance 2
Exhaust 32
Eye protection 76

Fairings 125
Fatigue failure 117
Fire precautions 78–83
Firing order 19

Flux 146
Flywheel inertia 18
Force 254–255
Four-stroke 11–16
Frame tubing 118
Frames 123
Friction 56, 177, 256–257
Fuel 6
Fuel tap 48
Full-film lubrication 57

Gas flow 28
Gear type oil pump 60
Gearbox 264–271
Goggles 96
Golden Era 340
GRP 226

Hand protection 76
Health and safety 75
Helmet 96
Helmholtz theory 34
History of motorcycles 331
Horn 210
Horse power 1
Hoses 73
Hot rolling 112

Identification 2
Ignition 35
Ignition coil 37
Ignition switch 36
Ignition timing 42
Impact and momentum 257
Induction 31
Inspection, test re-build 299–323

Joule 3

Kilowatt 4

Lights 209
Liquid cooling 65
LNG 6
Locking fuel cap 49
Lubrication 55–74

Magneto 35
Manufacturing 215–218
Methanol 6
Motorcycle industry services 220–222

Motorcycle licences 95
Motorcycle shop 213–215
Motorsports valley 322

Newton's Laws 257

Octane number 26
Ohms 255–256
Oil filter 60
Oil level check 62
Oil pump 59
Oil seal 63

Pascal's law 181
Petrol injection 51–53
Pickling 112
Pinking 26
Piston rings 19
Post-ignition 26
Pre-ignition 26
Pressure 254–255
Pressure relief valve 59
Profit and loss account 245–246
Protective clothing 76

Radial brakes 193
Radially spoked wheel 160
Radiator 65, 68
Radiator pressure cap 71
Retailing 218–220
Reynolds tubing 118–119
RIDDOR 86
Running gear 159
Running-on 26

Sankey diagram 74
Schrader valve 168
Scooter monocoque 130
Screw thread 320
Shock absorber 164
SI system 249–251
Simple carburetter 49
Single flare 190
Skin fusion 143
Solder 144–146
Spark plugs 38–39
Starter motor 202–203
Steel 112
Sump 10
Surface fusion 143
Sustainability 325

Temperature and heat 253–254
Tempering 113
Thermostat 70
Timeline 345–351
Torque 2
Total fusion 143
Trade associations 222
Tuning 275–298
Two-stroke 16–18
Two-stroke lubrication 64
Types of oil 58
Tyre parts 167

Valve cover 10
Valves 21
VIN number 2
Viscosity 57
Volts 255–256
Volumetric efficiency 23

Watts 255–256
Wheel bearings 165
Work done 2
Work hardening 117

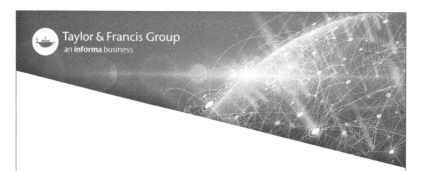

For Product Safety Concerns and Information please contact our EU
representative GPSR@taylorandfrancis.com
Taylor & Francis Verlag GmbH, Kaufingerstraße 24, 80331 München, Germany

www.ingramcontent.com/pod-product-compliance
Ingram Content Group UK Ltd.
Pitfield, Milton Keynes, MK11 3LW, UK
UKHW021450080625
459435UK00012B/445